P9-BBO-882

A NATURAL HISTORY OF THE FUTURE

ALSO BY ROB DUNN:

Delicious

Never Home Alone

Never Out of Season

The Man Who Touched His Own Heart

The Wild Life of Our Bodies

Every Living Thing

A NATURAL HISTORY OF THE FUTURE

What the Laws of Biology Tell Us About the Destiny of the Human Species

ROB DUNN

BASIC BOOKS
New York

Basic Books
Hachette Book Group
1290 Avenue of the Americas, New York, NY 10104
www.basicbooks.com
Printed in the United States of America

First Edition: November 2021

Published by Basic Books, an imprint of Perseus Books, LLC, a subsidiary of Hachette Book Group, Inc. The Basic Books name and logo is a trademark of the Hachette Book Group.

Print book interior design by Jeff Williams

Library of Congress Cataloging-in-Publication Data
Names: Dunn, Rob R., author.
Title: A natural history of the future : what the laws of biology tell us about the destiny of the human species / Rob Dunn.
Description: First edition. | New York : Basic Books, 2021. | Includes bibliographical references and index.
Identifiers: LCCN 2021012189 | ISBN 9781541619302 (hardcover) | ISBN 9781541619296 (ebook)
Subjects: LCSH: Nature—Effect of human beings on. | Human ecology. | Environmental sociology. | Ecological forecasting. | Environmental policy.
Classification: LCC GF75 .D86 2021 | DDC 304.2—dc23
LC record available at https://lccn.loc.gov/2021012189

ISBNs: 978-1-5416-1930-2 (hardcover), 978-1-5416-1929-6 (ebook)

TR

10 9 8 7 6 5 4 3 2 1

To my father,
who always likes to have a plan.

Contents

Introduction

I GREW UP LISTENING TO STORIES OF RIVERS. IN THE STORIES, HUMANS confronted the rivers. In the stories, the river always won.

In my childhood the rivers were the Mississippi River and its tributaries. I grew up in Michigan, but my father's father's family was from the town of Greenville, Mississippi. The Greenville of my grandfather's childhood was located on the ancient floodplain behind the earthen levee meant to hold the Mississippi River back. The Mississippi River could swallow boats. It would swallow small boys. And when my grandfather was about nine years old, it swallowed the entire town of Greenville. Houses floated downriver. Cows were strangled on their ropes as the river pulled them away. Many hundreds of people drowned. The town was never the same afterward.

The Great Flood, which occurred in 1927, was the sort of disaster that seemed to demand an explanation. The explanation depended on who was telling the story. One version blamed "gentlemen" from Arkansas, Mississippi's cross-river neighbor to the west. If the levee holding back the river on the Mississippi side were to break, the water would inundate Mississippi and spare Arkansas, which is just what happened during the Great Flood. Hence, some say (with no evidence whatsoever) that a group of gentlemen from Arkansas took their boats across the river and used dynamite to blow a hole in the levee and flood Greenville. In other versions, the flood was brought forth as the punishment of an angry God. Floodwaters and plagues have been among the favorite tools of vengeful gods going back to the earliest recorded Sumerian stories. In the version of the story I remember hearing most often, the water simply got too high and eventually began to make the levee bubble and then liquify. In some of the retellings, my grandfather was the boy who spotted the place where the levee began to liquify and notified people in town.

The truest story about the flood of Greenville is that it was caused by human attempts to control the river. It is in the nature of rivers to meander beyond their banks, carving new courses across the landscape. But a meandering river was and is ill-suited for houses, let alone cities, built near the river. It was and is ill-suited for big ports built along the river. In the years leading up to the Great Flood, the people living along the river spent inordinate amounts of money building levees to keep the river from meandering. The course of the river, previously governed by time, physics, and chance, was made artificial. It would be said that it was "tamed," "controlled," and even "civilized" so as to allow cities to grow and wealth to accumulate. The taming of the river was carried out with a sense of pride and, at times, hubris. It was the hubris associated with a belief in the ability of humans to take nature and bend it to a more human design.

For millions of years the Mississippi spilled over its banks each year, flooding the flat plains alongside the river. And it meandered, moving this way and that, creating new habitat and even new land as it did. As Amitav Ghosh noted in *The Great Derangement*, about the Bengal Delta, "the flow of water and silt [was] such that geological processes that usually unfold in deep time appear[ed] to occur at a speed where they [could] be followed from week to week and month to month."[1] The geography of Louisiana, for instance, is the consequence of the river's ancient movements; the state is the mouth of the river that drains a continent.

Trees evolved to rely on the floods and movements of the rivers, as did grasses. Fish relied on this swamping exuberance of water as part of their natural cycle of life and death. Native Americans along the Mississippi timed their farming, foraging, and ceremonies to these cycles and built settlements on ground high enough to escape the water when necessary. Nature and Native Americans alike responded to the river by working with it, taking advantage of its inevitable seasons and episodes. But the large-scale commercial transport along the Mississippi that fed early industrialization could not wait on nature and could not be bothered with its seasons or chronic movements. The early days of American industrialization required that boats travel on regular schedules and that cities, the ultimate destinations of the boats' goods, be as close as possible to the river. Industrialization required the river to be consistent, not just predictable.

Attempts to make the river consistent were attempts to make the river part of the broader realm of human control. Its banks were talked about as though they were analogous to pipes through which water flowed and could be redirected, slowed, sped up, or even stopped. The consequences of this view of the river were many. The consequences flooded my grandfather's home. The river was still wild. The river is still wild. Regardless of our interventions, the river, as the poet A. R. Ammons put it, will "go on with the ongoing."[2]

Even now that it is even more restrained, the Mississippi River will continue, every so often, to eat boats, small boys, and farms. It will flood towns, and we will be somehow surprised when it does. These floods will grow worse because of climate change. The river's predations are a reminder that nature will devour human attempts to escape, combat, or dominate nature. In this, the Mississippi River is like the river of life of which we are part. Our attempts to control the Mississippi are a metaphor for our attempts to control nature in general, but especially for our attempts to control life.

When we imagine the future, it is common to imagine ourselves nested within an ecosystem of technology, an ecosystem populated by robots, devices, and virtual realities. The future is shining and technological. The future is digital, ones and zeros, electricity and invisible connections. The dangers of the future—automation and artificial intelligence—are, as a slew of new books have pointed out, of our own invention. Nature is an afterthought in our contemplation of what comes next, a transgenic potted plant behind a window that does not open. Most depictions of the future do not even include nonhuman life, except on distant farms (tended by robots) or in indoor gardens.

We imagine a future in which we are the only living protagonists. We seek, collectively, to simplify the living world and channel it into our service, to circumscribe it so fully within our powers that it ceases to even be visible. We put up a levee between our civilizations and the rest of life. That levee is a mistake, both because it is not possible to hold life at bay and because in trying to achieve such a scenario, we do so at our own expense. It is a mistake with regard both to our place in nature and to what we know about the rules of nature and of the human relationship with the rest of nature.

We are taught some of the laws of nature in school. We know about gravity, inertia, and entropy, to name a few. But these are not the only laws of nature. Beginning with Charles Darwin, biologists have discovered, as the writer Jonathan Weiner put it, "laws of terrestrial motion as simple and universal as the physicists',"[3] laws of the motions of cells, bodies, ecosystems, and even minds. These are the biological laws that we need to have in the front of our mind if we are to make any sense of the years ahead. This is a book about those laws and what they tell us about the natural history of the future.

Some of the laws of biological nature, the kind I've most often studied, are laws of ecology. The most useful laws of ecology (and related fields such as biogeography, macroecology, and evolutionary biology) are, like the laws of physics, universal. These biological laws of nature, like the laws of physics, allow us to make predictions. However, as physicists have pointed out, they are more limited than the laws of physics because they only apply to the tiny corner of the universe in which life is known to exist. Still, given that any story that involves us also involves life, they are universal relative to any world we might experience.

It is easy to get caught up in whether to call the rules of biological nature "laws," as I do here, "regularities," or something different. I'll leave that debate to the philosophers of science. In keeping with the everyday usage of the term, I will call them "laws." These are the "laws of the jungle"—or, rather, the laws of the jungle, prairie, swamp, and, because our homes are also alive, bedroom and bathroom. Ultimately, I am most concerned with the reality that knowing about such laws helps us understand the future into which we are—arms flailing, coal burning, and full speed ahead—hurling ourselves.

Most of the laws of nature are, to ecologists, well known. Most of them were first studied more than a hundred years ago and have been elaborated and refined in recent decades with advances in statistics, modeling, experiments, and genetics. Because these laws are known and intuitive to ecologists, ecologists often don't mention them. "Of course that is true. Everyone knows. Why talk about it even?" But these laws are often not intuitive if you haven't spent recent decades thinking and talking about them. And what is more, when the future is considered, these laws nearly all lead to conclusions and consequences that surprise even ecologists, conclusions and consequences at odds with many of the decisions we make in our everyday lives.

One of the most robust biological laws is natural selection. Natural selection is Charles Darwin's elegant revelation of the way life evolves. Darwin chose the term "natural selection" to reflect the reality that in each generation, nature "selects" some individuals relative to others. It selects and disfavors those individuals with traits that make them less likely to survive and reproduce. It selects and favors those individuals with traits that make them more likely to survive and reproduce. The favored individuals are the ones that pass on their genes and the traits those genes encode.

Darwin imagined natural selection to be a slow process. We now know that it can happen very quickly. Evolution by natural selection has been observed in real time in many, many species. None of this is surprising. What is surprising is the river-like inevitability with which the consequences of this simple law flow into our daily lives each time we, for example, try to kill a species.

We try to kill species when we use antibiotics, pesticides, herbicides, and any other "-cide." We do this in our homes, hospitals, backyards, farm fields, and even, in some cases, forests. When we do, we are attempting to exert control in much the same way as those who built the levees along the Mississippi River also tried to exert control. The effects are predictable.

Recently, Michael Baym and colleagues at Harvard University constructed a giant Petri plate, a "megaplate," divided into a series of columns. I feature this megaplate and its columns in Chapter 10. It is a plate of great significance. Into the megaplate Baym put agar, which is both food and habitat for microbes. The outside column on each side of the megaplate contained agar and nothing more. Moving inward, each subsequent column was laced with antibiotics at ever-higher concentrations. Baym then released bacteria at both ends of the megaplate to test whether they would evolve resistance to the antibiotics.

The bacteria had no genes that conferred resistance to the antibiotics; they entered the megaplate as defenseless as sheep. And if the agar was the pasture for these bacterial "sheep," the antibiotics were the wolves. The experiment mimicked the way we use antibiotics to control disease-causing bacteria in our bodies. It mimicked the way we use herbicides to control weeds in our lawn. It mimicked each of the ways we try to hold back nature each time it flows into our lives.

So what happened? The law of natural selection would predict that so long as genetic variation could emerge, via mutation, the bacteria should eventually be able to evolve resistance to the antibiotics. But it might take years or longer. It might take so long that the bacteria would run out of food before they evolved the ability to spread into the columns with antibiotics, the columns filled with wolves.

It didn't take years. It took ten or twelve days.

Baym repeated the experiment again and again. It played out the same each time. The bacteria filled the first column and then briefly slowed, before one and then many lineages evolved resistance to the lowest concentration of the antibiotic. Those lineages then filled that column and slowed again, briefly, before another lineage and then, again, many lineages evolved resistance to the next highest concentration of antibiotics. This continued until a few of the lineages evolved resistance to the highest concentration of antibiotics and poured into the final column, like water over a levee.

Seen sped up, Baym's experiment is horrifying. It is also beautiful. Its horror lies in the speed with which bacteria can go from being defenseless to indestructible relative to our power. Its beauty lies in the predictability of the experimental results, given an understanding of the law of natural selection. This predictability allows two things. It allows us to know when resistance might be expected to evolve, whether among bacteria, bedbugs, or some other group of organisms. It also allows us to manage the river of life so as to make the evolution of resistance less likely. An understanding of the law of natural selection is key to human health and well-being and, frankly, to the survival of our species.

There are other biological laws of nature whose consequences are similar to those of natural selection. The species-area law governs how many species live on a particular island or habitat as a function of its size. This law allows us to predict where and when species will go extinct, but also where and when they will evolve anew. The law of corridors governs which species will move in the future as climate changes, and how. The law of escape describes the ways in which species thrive when they escape their pests and parasites. Escape accounts for some of the successes of humans relative to other species and for how we have been able to achieve such extraordinary abundance relative to other species. The law frames some of the challenges that we will face in the coming years when our possibilities of escape (from pests, parasites, and the like) are fewer. The law of the niche governs where species, including humans, can live and where we are likely to be able to successfully live in the future as climate changes.

These biological laws are alike in that their consequences play out independently of whether or not we pay them heed. And, in many cases, our failure to pay them heed ushers us into trouble. Failing to pay attention to the law of corridors leads us to inadvertently help problem species (rather than beneficial or simply benign species) into the future. Failure to pay attention to the species-area law leads to the evolution of problem species such as a new species of mosquito in the London Underground railway system. Failure to pay attention to the law of escape leads us to squander moments and contexts in which our bodies and crops are free of parasites and pests. And so on. Conversely, the laws are also similar in that if we pay them heed, if we consider how they will influence the natural history of the future, we can create a world that is more forgiving of our own existence.

Other laws relate to the ways in which we, as humans, behave. As laws of human behavior they are both narrower and messier than the broader laws of biology; they are as much tendencies as laws. Yet they are tendencies repeated across times and cultures, tendencies that are relevant to understanding the future both because they suggest how we are most likely to behave and because they also indicate what we need to be aware of if we are to go against the rule.

One of the laws of human behavior relates to control, to our tendency to try to simplify life's complexities, just as one might try to straighten and channel an ancient and powerful river. The coming years will present more novel ecological conditions than have occurred in millions of years gone by. Our human populations will swell. More than half the Earth is now covered by ecosystems we have created—cities, farm fields, waste-treatment plants. We now, meanwhile, control, directly and incompetently, many of the most important ecological processes on Earth. Humans now eat half of all the net primary productivity, the green life that grows, on Earth. And then there is the climate. In the next twenty years, climatic conditions will emerge unlike any humans have ever been exposed to before. Even under the most optimistic scenarios, by the year 2080 hundreds of millions of species will need to migrate to new regions and even new continents in order to survive. We are reshaping nature at unprecedented scales, and for the most part, we are absentmindedly looking the other way while doing so.

As we reshape nature, our behavioral tendency is to use more and more control: to make our farm fields simpler and more industrial and, to return to a previous example, to make our biocides ever stronger. This, I will argue, is a problematic approach in general, but it is especially so in a changing world. In a changing world, our behavioral tendency to try to control is at odds with two laws of diversity.

The first law of diversity is manifest in the brains of birds and mammals. In recent years, ecologists have revealed that animals with brains capable of using inventive intelligence to carry out novel tasks are favored by variable environments. These animals include crows, ravens, parrots, and some primates. Such animals use their intelligence to buffer the diverse conditions they encounter, a phenomenon described as the law of cognitive buffering. When environments that were once consistent and stable become variable, these species with inventive intelligence become more common. The world becomes a crow's world.

A second law of diversity, the diversity-stability law, states that ecosystems that include more species are more stable through time. An understanding of this law and of the value of diversity proves useful in the context of agriculture. Regions with a greater diversity of crops have the potential to have more stable crop yield from year to year and hence less risk of crop shortages. Repeatedly, although our tendency is often to try to simplify nature when we are confronted with change, or even to rebuild it from scratch, maintaining nature's diversity is more likely to lead to sustained success.

When we try to control nature, we often come to imagine ourselves as outside nature. We speak of ourselves as if we were no longer animals, as if we were a species alone, disconnected from the rest of life and subject to different rules. This is a mistake. We are both part of and intimately dependent on nature. The law of dependence states that all species depend on other species. And we, as humans, are probably dependent on more species than any other species ever to exist. Meanwhile, just because we depend on other species does not mean nature depends on us. Long after we go extinct, the rules of life will continue. Indeed, the worst assaults we carry out on the world around us nonetheless favor some species. What is remarkable about the big story of life is the extent to which it is ultimately independent of us.

Finally, one of the most consequential sets of laws regulating how we plan for the future relates simultaneously to our ignorance about nature and our misperceptions about its dimensions. The law of anthropocentrism states that, as humans, we tend to imagine the biological world to be filled with species like us, species with eyes, brains, and backbones. This law emerges from the limits of our perception and the limits of our imaginations. It is possible that we might someday escape this law and break through our ancient biases—possible but, for reasons that I elaborate, unlikely.

A decade ago, I wrote a book called *Every Living Thing* about the diversity of life and what remains to be discovered. Life, I argued in that book, is far more diverse and ubiquitous than we imagine. The book was an extended consideration of what I call Erwin's law.

Repeatedly scientists have announced the end (or near end) of science, the discovery of new species, or the discovery of life's extremes. Usually, in doing so, they position themselves as having been key to putting the final pieces in place. "Finally, now that I am done, we are done. Look what *I* know!" And repeatedly, after such announcements, new discoveries have revealed life to be far grander and more poorly studied than had been imagined. Erwin's law reflects the reality that most of life is not yet named, much less studied. Erwin's law is named for a beetle biologist, Terry Erwin, who, with a single study in a rain forest in Panama, changed our understanding of the dimensions of life. Erwin initiated a revolution in our understanding of life analogous to the Copernican Revolution. The Copernican Revolution was complete when scientists came to agree that Earth and the other planets circled the sun. The Erwinian revolution will be complete when we remember that the living world is far vaster and more unexplored than we imagine it to be.

Collectively, these laws of the living world and our place in it offer a vision for what is and is not possible with regard to the natural history of the future and our place in it. Only by keeping life's laws in mind can we imagine a sustainable future for our species, a future in which our cities and towns are not flooded again and again by the consequences of our failed attempts to manage life—flooded not only by water but also by pests, parasites, and hunger. We will fail again and again if we ignore these laws. The bad news is that our default approach to nature seems to be to try to hold it back. We tend to fight nature at our own expense and then blame vengeful gods (or gentlemen from Arkansas) when things don't work out. The good news is that it doesn't have to be that way: if we pay attention to a set of relatively simple laws of life, we have a much better chance at surviving a hundred years, a thousand years, or even a million years. And if we don't, well, ecologists and evolutionary biologists together actually have a pretty good idea of the trajectory of life in our absence.[4]

CHAPTER 1

Blindsided by Life

THE FIRST SPECIES OF HUMAN, *HOMO HABILIS*, EVOLVED ROUGHLY 2.3 million years ago. *Homo habilis* then begat *Homo erectus*. *Homo erectus*, in turn, begat a dozen or so other human species, including, eventually, Neanderthals, Denisovans, and *Homo sapiens*. All of this transpired over years during which many mammal species were very numerous. Reindeer numbered in the millions. Some mammoth species numbered in the hundreds of thousands. Yet the largest population ever attained by any human species between 2.5 million years ago and 50,000 years ago would have been around ten to twenty thousand individuals. These individuals would have been organized into highly dispersed, relatively small groups. At no time and nowhere did they abound. For essentially the entirety of prehistory, humans were relatively rare, their survival far from inevitable. That would change.

Around fourteen thousand years ago, our species, *Homo sapiens*, began to settle into more sedentary lives. For some populations, hunting and gathering gave way to farming, beer brewing, and baking. This transition brought population growth, which continued over the succeeding millennia. Roughly nine thousand years ago, as the first small cities began to emerge, the total number of humans on Earth was still relatively small, and yet the rate of human population growth had begun to increase. By the year zero, the total population on Earth may have been ten million, which is to say, the size of a modern Chinese city of no special renown. Yet, the rate of human population growth was continuing to increase.

Then, between the year zero and today, that rate accelerated. Earth added eight billion people. This increase in human populations has been called "the great escalation" or the "great acceleration." The consequences of humans escalated, and the rate at which those consequences increased, year by year, accelerated.[1]

In the laboratory, we see the kind of population growth that humans underwent during the great acceleration when we study bacteria and yeasts. A few small settlement-like colonies on a Petri dish, when given as much food as they want and need, initially grow slowly, but growth accelerates until the food is devoured and the Petri dish is covered with bubbling life. We are that bubbling life on Earth's Petri dish, a reality that began to be noticed as early as 1778 when the French naturalist, Georges-Louis Leclerc, the comte de Buffon, wrote, "The entire face of the Earth bears the imprint of human power."[2]

During the great acceleration, the proportion of Earth's biomass consumed by humans increased exponentially until, today, more than half of all the green growth on Earth, the terrestrial primary productivity, is consumed by humans. By one estimate, 32 percent of the terrestrial vertebrate biomass on Earth is now composed of nothing more than fleshy, human bodies. Domestic animals make up 65 percent. Just 3 percent is left over for the rest of vertebrate life, the remaining tens of thousands of boney animal species. Unsurprisingly in this context, rates of extinction have increased more than a hundredfold, perhaps much more. Any measure of the human effect on life over the last twelve thousand years shows a line rising, often exponentially. It is true of the pollutants produced by human societies. Methane emissions have increased by 150 percent. Nitrous oxide emissions have increased by 63 percent. Carbon dioxide emissions have nearly doubled to levels last seen three million years ago. The trends are similar for pesticides, fungicides, and herbicides. These effects are all increasing, all accelerating in step with the growth of our populations, needs, and desires.

At some hard-to-define point during the great acceleration, the populations and actions of humans ushered in a new geological epoch, the Anthropocene. It all happened so quickly. Compared to the long history of life, the growth of human populations was instantaneous. A train crash. An explosion. A mushroom rising from the wet ground of our origin. In confronting the consequences of this rise, as if studying the aftermath of a collision, one gathers the pieces and imagines that if enough pieces, enough details, are gathered, the whole will make sense. This seems to be a logical supposition, so logical that it has become a common approach to doing science. For biologists, the pieces that are gathered are species. Biologists examine the species. Biologists chart their details and their needs. But there is a problem with this approach: our own lack of awareness.

The species we study to understand the world are nearly all unusual species. They are species that are representative neither of the realities of the living world nor of the portion of the living world most likely to affect our own well-being. Our problem is simple. We tend to assume the living world to be both like us and relatively well understood. Both of these assumptions are wrong, the result of law-like biases in the way we make sense of the world. I begin by considering these biases because we can't understand the natural history of the future without being aware of the wide gulf between our perceptions of the biological world and its more interesting realities.

The first of our biases is anthropocentrism. This bias is so deeply part of our senses and psyche that it might be called a law, the law of anthropocentrism. The law of anthropocentrism is grounded in our biology. Every animal species has a perception of the world framed by its own senses. If it were dogs that were in charge of science, I'd be writing about the problem of caninecentrism. But what is unique with humans is that our bias influences not only the way we individually perceive the living world around us, but also the scientific system we have built to catalog the world. It was the Swedish natural historian Carl Linnæus who gave our system its rules, but he also gave the system's anthropocentrism momentum, inertia, and a peculiar geography.

Linnæus was born in 1707 in the village of Råshult, about 150 kilometers northeast of the city of Malmö in southern Sweden. Råshult has a climate more or less like that of Copenhagen, Denmark. It has some of the coldest summers in the world, and its winters are sufficiently dark and cloudy that when the sun appears, people turn their faces to it like sunflowers. They even point. "There it is!" It was in Råshult that Linnæus became interested in nature; it was farther north in Sweden, in Uppsala and its surrounds, that he would study nature.

Sweden, despite its large size, is among the least biologically diverse countries in the world. Yet Linnæus assumed that the biological poverty of his home place was the norm. Linnæus's

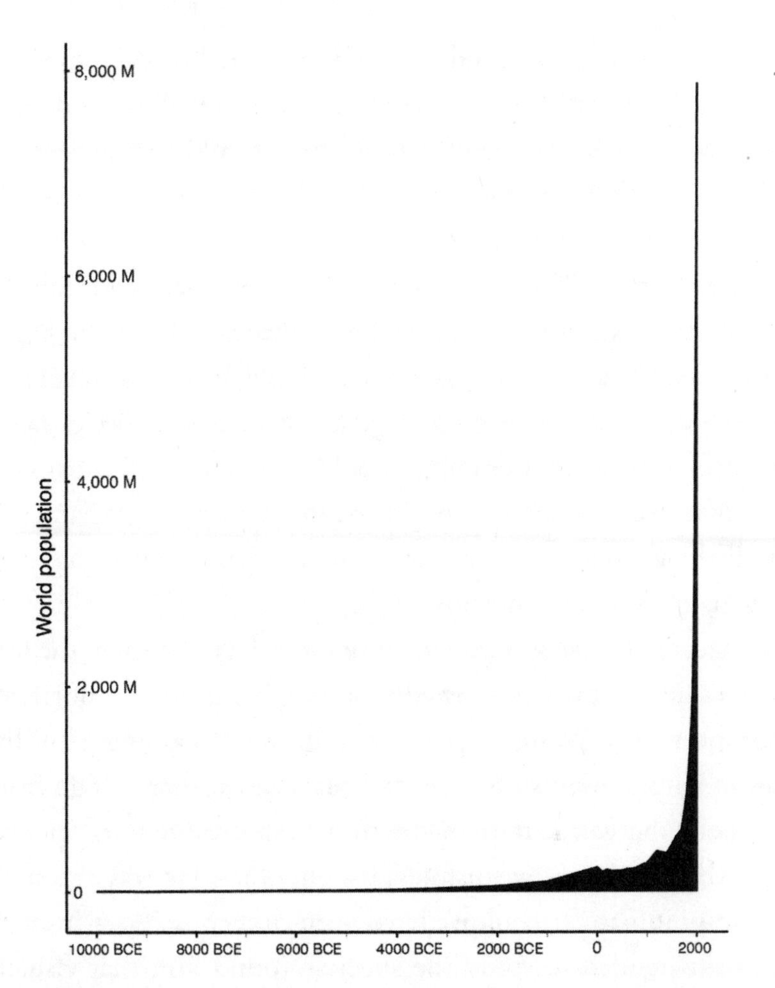

Figure 1.1. Population growth of humans over the last twelve thousand years. Prior to twelve thousand years ago, that is, before 10,000 BCE, human populations are thought to have never numbered more than about 100,000 globally, a tally that would not show up on this graph. Figure by Lauren Nichols.

trips outside Sweden were to the Netherlands, northern France, northern Germany, and England. These regions are slightly more southerly than Sweden, and yet, relatively speaking, they are much the same in terms of their biology. As seen and imagined by Linnæus, Earth's landscape was, if not uniformly Swedish, at least

Swede-*ish*. It was rainy and cold and populated by deer, mosquitoes, and biting flies and by beech, oak, aspen, willow, and birch trees. It was a landscape of delicate spring flowers, late summer berries, and fungi pressing up out of the ground in the wet fall, just in time to be eaten.

Before the 1700s, scientists in different places and cultures had different systems for naming life. Linnæus codified and began to implement a universal system, a scientific common tongue, in which each species was given a genus name and a species name in Latin; humans, for example, would be *Homo* (our genus) *sapiens* (our species). He then considered the species near at hand. He studied and touched them, bestowing on them, as if in blessing, new names—Linnaean names.

Because Linnæus began renaming species in Sweden, the first species he renamed were Swedish and, more generally, northern European. The Western scientific tradition of naming all of life began with a Swedish bias. Even today, the farther you go from Sweden, the easier it is to discover a species new to science. Nor was Linnæus's Swedishness his only bias. He was also inescapably human. It couldn't have been otherwise. As a human, Linnæus tended to study the species around him that visually commanded his attention. Linnæus liked plants and had a particular fascination with their sex parts. But he also studied animals. Within the animal kingdom, vertebrates received most of his focus. Among the vertebrates, Linnæus tended to pay attention to mammals. Within mammals, Linnæus tended to ignore small species, such as the innumerable kinds of mice, preferring to feature bigger species. In general, his focus was either on species that were visually pleasing or obvious to him and to his colleagues, such as flowering plants, or on species that were enough like us in size or behavior to be both easily seen and relatable.

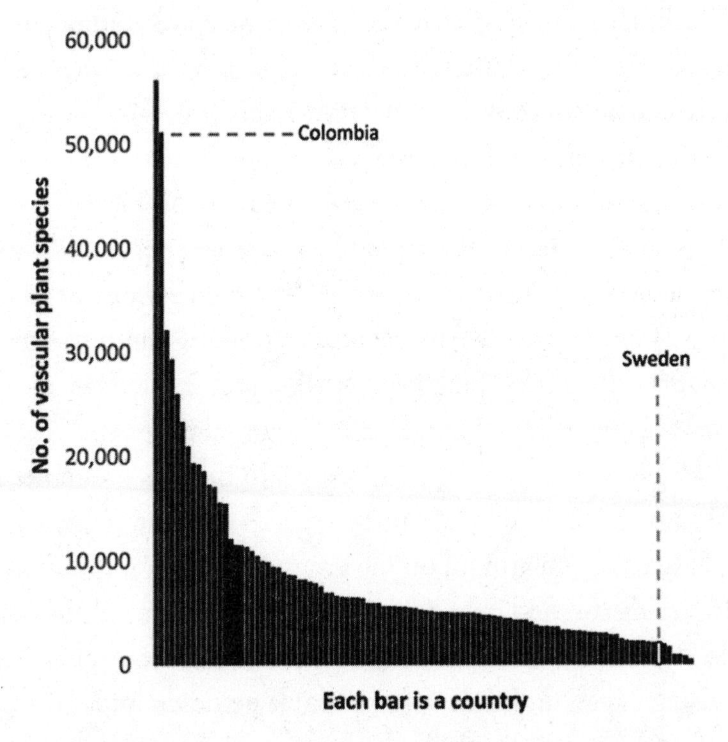

Figure 1.2. The number of vascular plant species in each of 103 different countries. Note that Sweden is among the least biologically diverse countries with regard to plant diversity. For example, although Colombia is just twice as big as Sweden, it contains roughly twenty times as many plant species as does Sweden. Patterns in the diversity of, for example, birds, mammals, and insects are similar.

In this way, his focus was both Eurocentric and anthropocentric. The scientists that Linnæus trained, and modestly called his "apostles," for the most part followed in his footsteps and had similar biases. So too have most scientists since. These biases influence not only which species are named first,[3] but also which species are studied in detail and especially which species are the subject of conservation efforts.

The problem with the Eurocentric and anthropocentric biases of science are that they give us a false impression of the world.

They lead us to imagine that the species we have studied are a reflection of the world itself, rather than just the part of the world we have chosen to study. Several decades ago, it became clear just how wrong this perception was when scientists started to consider a simple question: "How many species are there on Earth?"

Attempts to answer this question in earnest began with the entomologist Terry Erwin. In the 1970s, Erwin set out to study a group of beetles that live in the tops of tropical rain forest trees in Panama. These tree-living beetles, which most often live at the interface between branches and clouds, are called *ground* beetles because they were first studied in Europe. In Europe, ground beetles are not terribly diverse, but the species that are present do, indeed, tend to run around on the ground.

In trying to find and identify ground beetles in the sky, Erwin deployed a new method. He would climb up into a tall tree, using ropes, and then spray a fog of pesticide into the canopy of an adjacent tree. Initially, he fogged trees of the species *Luehea seemannii*. After fogging trees, he returned to the ground and then waited for dead insects to fall. When Erwin tried this method for the first time, the insects fell by the tens of thousands onto tarps that he had stretched out on the forest floor. To his delight, there were ground beetles, but there was also a great deal more.

Erwin would ultimately tally about 950 species of beetles in *Luehea seemannii* trees, at least for the kinds of beetles that he and his collaborators could identify. On top of that, he estimated there were an additional 206 species of beetles in his samples from the weevil family, though no weevil expert had time to formally make the necessary identifications. The resulting total of about 1,200 species of beetles amounts to more beetle species in one kind of tree in one forest than there are bird species in the

United States. Erwin next considered other kinds of insects and then other kinds of arthropods more generally. He came to notice that not only were most of the species of ground beetles new to science, but so too were most of the species of other kinds of beetles and most of the species of each and every other kind of arthropod. What was more, when Erwin started sampling other kinds of trees, he saw different species than he had on the *Luehea seemannii* trees. Each rain forest tree species had its own insect and other arthropod species, and tropical rain forest tree species are extraordinarily diverse.

Erwin was confronted with a riot of unnamed life. He was surrounded by species no scientist had ever seen before, much less studied in any detail. No one knew anything about these species, other than the trees from which they had fallen. It was at this point that Erwin received a call from the botanist Peter Raven. Raven, then director of the Missouri Botanical Garden, asked Erwin a simple question. If there were so many unnamed beetle species in a single tree of a single species, "how many species might there be in an entire acre of forest in Panama?" Raven's question was prompted by work he was doing as chair of a National Research Council committee charged with identifying the gaps in our understanding of tropical forest biology.[4] Erwin responded, "Peter, nobody knows that stuff about insects. It's just impossible."[5]

AT THE TIME Raven called Erwin, there was no good estimate of the diversity of life on Earth. In 1833, the entomologist John Obadiah Westwood polled his entomological acquaintances and, on the basis of the results, hypothesized that there might be five hundred thousand insect species on Earth, to say nothing of other kinds of organisms. In the context of his report to the National Science Foundation, Raven had also offered an estimate, based on some simple math. He predicted there might be three to four million species on Earth. If Raven was right, more than half of all species on Earth were unnamed.

Meanwhile, although Erwin had said it was "impossible" to estimate the number of species of insects in an acre of forest in Panama, much less the number of all species on Earth, he decided to give it a try. He started by doing some calculations. If there were 1,200 species of beetle in the *Luehea seemannii* trees, and one-fifth of those beetle species were dependent on that particular tree species, how many beetle species might there be in a hectare of Panamanian forest? Assuming that the discoveries he had made in *Luehea seemannii* trees were representative of the sort of specialization he might find on other tropical trees, Erwin calculated the number of beetle species in a Panamanian forest, given the number of tree species present. He then adjusted his figures to get an estimate of the total number of arthropods (encompassing not just insects, but also spiders, centipedes, and the like) more generally. The result was forty-six thousand species of arthropods in a hectare of forest in Panama. That was his answer for Raven (though it was a little late—Raven's report to the National Science Foundation had, by then, been long since published). But Erwin decided to go a little further. He used the same sort of simple math to estimate the number of arthropod species not just in a hectare of forest in Panama or all the forests in Panama but, instead, in all the tropical forests of the world. If there were about fifty thousand tropical tree species on Earth, Erwin wrote in a two-page paper in the *Coleopterists Bulletin*, "there might be 30 million tropical arthropod species in the world." Given that only about a million species of arthropods (and 1.5 million species of organisms more generally) were named at the time, this would mean that nineteen out of every twenty species of arthropods were not yet named![6]

Erwin's estimate provoked a wave of academic controversy. Scientists debated its validity aggressively in print and passive-aggressively in person.

Some scientists suggested, in private, that Erwin was foolish. Some said it in public. Some thought him foolish because his estimate was too high. Others thought he was foolish because his estimates for their own favorite groups of organisms were too low. Dozens of scientific papers were written. Erwin wrote responses to the responses to his papers. He collected new data. He wrote more papers, which, in turn, triggered new responses. Meanwhile, other scientists were inspired to collect new data. More papers were written. The work of refining, rejecting, or improving Erwin's estimate was aggressive, furious, contested, and public.

Eventually, the debate basically ceased, or at least slowed dramatically. After years of debate, scientists had reached a kind of quiet consensus; the number of unnamed species of animals was sufficiently large that it would be centuries before we know for sure whether Erwin was right. The most recent estimate of the number of insect and other arthropod species on Earth suggested there might be about eight million, which is to say that seven out of eight animal species are not yet named. Eight million is fewer species than Erwin hypothesized and yet still far more than had ever been imagined before his work.[7] The unknown is large; the known is humble.

In causing scientists to reconsider the dimensions of animal life, Erwin served as a kind of Copernicus of biodiversity. The astronomer Copernicus argued that the universe was heliocentric. Earth, Copernicus said, circled the sun rather than the other way around, and in addition, Earth rotated on its axis once a day. Erwin, meanwhile, revealed us to be just one animal species among millions. He also revealed that the average animal species is not a vertebrate like us, or northern (like Linnæus). It is instead a tropical beetle, moth, wasp, or fly. Erwin's insights were radical. Indeed, they were so radical that it has proven more difficult to incorporate them into our daily understanding of the world than it was to imagine that still-seeming Earth is both spinning on its axis and circling the sun.

The Erwinian revolution in our perspective does not end with insects. Fungi, such as those that produce mushrooms, appear to be even more poorly known than are insects. My colleagues and I recently studied the fungi found inside houses across North America. We found fungi in every house. But what was remarkable was not the presence of fungi but, instead, the number of kinds of fungi. The most recent tallies of all the named fungi in North America noted roughly twenty thousand species. By studying the dust in houses, we found twice as many species.[8] That is to say that no fewer than half of those species we found in houses must be new to science—thousands of fungus species new to science in homes. It isn't that houses are special. Instead, the teeming, unnamed, fungal multitudes in our own homes simply point to our broader ignorance of the fungal life around us. Each time you breathe in, half of the kinds of fungal spores you inhale are yet to be named, much less studied in sufficient detail to understand their consequences for our own health and well-being. Pause now to take a breath; inhale the fungal unknown. Fungi are probably not as diverse as insects, but they are far more diverse than are vertebrates.

But it isn't the fungi that we must make sense of if we are to complete the Erwinian revolution; it is, instead, the bacteria. Linnæus knew of the existence of bacteria, but he ignored them. He lumped all microscopic life into what was, in effect, a single species, "chaos," too small and different to be organized or even organizable. Recently Kenneth Locey and one of my collaborators, Jay Lennon, tried to take the measure of this chaos. They focused just on bacteria and estimated that there might be a trillion kinds of bacteria on Earth. A trillion (1,000,000,000,000).[9] A *trillion*. Perhaps it was these multitudes that Terry Erwin had in mind when, later in his career, in a moment of humility before the grandeur, he noted that "biodiversity is infinite" and "there is no way to estimate the infinite."[10] Locey and Lennon's assessment of bacterial diversity was not that it is infinite, but relative to the known world, it is nearly so. Locey and Lennon based their estimate on the study of data from thirty-five thousand samples, from around the world, of soil, water, feces, leaves, foods, and other habitats in which bacteria dwell. In those samples, they were able to identify five million genetically different kinds of bacteria. They then used some of the general rules of life (for example, how the number of species in a habitat increases with the number of individuals in that habitat) to estimate how many kinds of bacteria they would have encountered were Earth to be sampled completely. The answer was a trillion, give or take a few billion. Locey and Lennon's estimate may well be very wrong, but it will be decades, maybe centuries, maybe longer, before we can say for sure. In a casual, drifty, late-day conversation, one of my close colleagues said she thought there were probably only a billion species of bacteria. But then she went on to say, "However, I have no idea. What I do know is that new bacteria species are everywhere." We are sitting on them, breathing them in, and drinking them; we are just not naming or counting them, or at least naming or counting them remotely fast enough to make sense of the wilderness we walk through each day.

By the time I was a graduate student, Erwin's estimate had led scientists to imagine that most species were insects. For a while, it seemed as though fungi might be the big story. Now it seems as though, to a first approximation, every species on Earth is a bacterial species. Our perception of the world keeps changing; more specifically, our measure of the dimensions of the biological world keeps expanding. And as it does, the average way of living in the world seems to be less and less like our own. The average animal species is not European, nor is it a vertebrate. And as for the average species more generally, it is neither animal nor vegetable; it is instead bacterial.

Bacteria, though, are not even the end of the story. Most individual strains and species of bacteria appear to have their own specialized viruses called bacteriophages. In some cases, as bacteriophage expert Brittany Leigh reminded me recently via email when reviewing this chapter, the number of kinds of bacteriophages outnumbers the number of kinds of bacteria ten to one. If there are a trillion species of bacteria, then it is possible there are also a trillion kinds of bacteriophages or even ten trillion kinds of bacteriophages. No one knows. What we do know, with certainty, is that the vast majority of species are not yet named or studied in any way or understood.

Beyond the bacteriophages, there is a final layer to this unraveling of our position at the center of things. It may be that the average species is not only not European, and not an animal, but also, not able to survive on the surface of the Earth, as I was recently reminded by Karen Lloyd, a microbiologist at the University of Tennessee.

Lloyd studies microbes that live beneath the ocean in Earth's crust. Not long ago, it was thought that Earth's crust was devoid of life. Research by Lloyd and others has shown, instead, that it is brimming with it. The organisms living in the crust do not depend on the sun for sustenance. They rely instead on energy generated by using gradients in chemicals deep down below us all. They use such energy in order to live simple, slothful lives.

Some of these organisms live so slowly that a single generation might take from a thousand to ten million years. Imagine, now, one cell of one of those latter, ten-million-year species. Imagine a cell that is about to divide, finally, tomorrow. It might have last divided before the ancestors of humans and gorillas began their separate trajectories. It would have last divided even before the ancestor of chimpanzees and humans diverged from the ancestor of gorillas. In one generation, such a cell would have lived through not only the entire sweeping evolutionary story of humans, but also all of the great acceleration. What will the next generation in that lineage experience in its lifetime, one that could conceivably end in about the year ten million?

These slow-living, chemical-eating crust microbes were discovered only relatively recently. But they are now thought to represent up to 20 percent of all of the living mass of life (what scientists call biomass) on Earth. Depending on how deep they go, this may be an underestimate. We have no idea how deep they go. Deeper, certainly, than we humans have been. The crust microbes aren't "normal." Theirs is not the average condition of life. Yet their lifestyle is actually more common, whether measured in terms of biomass or of diversity, than is the mammalian lifestyle or the vertebrate lifestyle.

The average species is neither like us nor dependent on us, in contrast to what our anthropocentrism would tend to suggest. This is the key insight of the Erwinian revolution that goes along with the recognition of what I call Erwin's law. Erwin's law states that life tends to be far less well studied than we imagine it to be. Together, the law of anthropocentrism and Erwin's law are hard to remember in our daily lives. It might require a kind of daily affirmation. "I am large in a world of small species. I am multicellular in a world of single-celled species. I have bones in a world of boneless species. I am named in a world of nameless species. Most of what is knowable is not yet known."

IT IS SURPRISING that we as a species have been as successful as we have despite our ignorance of the biological world and our biased perspective on its dimensions. Einstein said that "the eternal mystery of the world is its comprehensibility"; in other words, what is incomprehensible is how much we comprehend.[11] But I don't think that is quite right. I think that what is even more incomprehensible is that we have survived despite how little we have comprehended. We are like a driver who somehow gets down the road, despite being too short to see out the window, a little drunk, and very fond of acceleration.

In part we have been able to get by because we have been able to understand what the smaller, unnamed species around us do even if we don't know what they are. This, for example, is what bakers and brewers have long done in making sourdough bread or beer.

In making sourdough bread, one mixes flour and water, and then, over days, seemingly miraculously, the flour-water mixture begins to bubble and expand and become acidic. This bubbling mixture, called the starter, can then be added to more flour and water to make a dough that rises and becomes sour. The resulting sourdough can then be baked to yield bread. We don't know when the first sourdough bread was baked. I've recently begun collaborating with archaeologists on a project to consider whether a seven-thousand-year-old bit of charred food is the oldest sourdough bread. We don't yet know whether that bit of food is ancient sourdough (it seems like it might be). But even if it isn't, it seems likely that the oldest sourdough bread, when discovered, will be at least that old.

The oldest beer so far discovered actually predates agriculture.[12] The process used to make that beer would have been very similar to the process of making sourdough bread. Grains are sprouted. Those sprouted (malted) grains are boiled and left until they begin to become sour and alcoholic.

In both ancient brewing and baking, traditional scientists improved their ability to make better products through trial and error. Bakers figured out, for instance, that some starter could be stored up, fed, and reused to make new doughs bubble. They figured out what conditions the starter liked. They treated the starter like a hard-to-describe, and yet very important, family member. Similarly, brewers figured out how to take some of the froth from the top of one beer and add it to another. That froth too was a kind of "animal."

What the bakers didn't understand was that the starters rose because of ancient yeasts and that they became sour because of ancient bacteria. What the brewers didn't understand was that the beers became alcoholic because of ancient yeasts and became sour because of ancient bacteria. What was more, neither bakers nor brewers understood that the bacteria in the bread and beer were coming from the grains they were growing and from their own bodies. Nor did they grasp that the yeasts in the bread and beer were coming from the bodies of wasps (the natural habitat of beer and bread yeasts). It was enough to know the steps necessary to keep conditions right for these microbes, the recipe for going about daily life in a world full of unknowns.

However, as our ancestors began to change the world around them, they inadvertently also altered the composition of species around them. When they did, their recipes for daily life sometimes failed to work. Bread didn't rise. Beer didn't brew. They couldn't say why. They gave up, moved, innovated, or found new things to make. We don't see much of a record of the failures that led to these transitions. We just see the transitions. Sometimes the archaeological record forgivingly glosses over our missteps, the way a photograph taken in dim light and from a great distance can hide some wrinkles and blemishes. It is likely, however, that as human populations grew and as ecological changes due to humans accelerated, the number of such failures of the ancient recipes for daily living increased.

MANY YEARS AGO, I read a story by a science writer in which he entered a cave with a guide and a group of fellow travelers. As the group entered the cave, bats began to fly out in large numbers. The writer could hear their movement and chittering and could even feel the wind from their many wings. "Don't worry," the guide announced, "the bats know exactly where you are thanks to their ability to echolocate. They see us in the dark!" As the guide turned to walk farther into the cave a bat, flying quickly out into the night, hit him in the face—hard.

What the guide did not know was that while bats do have amazing abilities to "see" in the dark via echolocation, they also use a detailed knowledge of landmarks and repeated routes to find their way, especially in caves. The bat was flying along a preferred route and suddenly encountered the guide, who was, according to its model of the world, not there. The bat was blindsided by the man and the man by the bat.

Many of our past successes have been in a world of fixed objects, a world of relative stability. We charted our way even without being able to see clearly. But by altering the life around us, we have created a situation like that faced by the bat. As we confront the future, our collective bearings are off, and our perception of the world around us is deeply flawed. Nothing is where it used to be. We have begun to crash into things; we find ourselves blindsided by life.

In some cases, the consequences of our stumbles are problematic but not deadly. Such cases offer a window into a broader fall. For example, my collaborators recently tried to make and study sourdough starters in a laboratory at North Carolina State University, a laboratory full of the same sorts of unusual microbial species common in homes that are sealed tight and where food is rarely fermented. When we did, the result was unsuccessful. Few yeasts colonized the starters. Instead, the starters were colonized by filamentous fungi known as molds; molds do not leaven bread. By bringing bread making into the laboratory, we had altered some component of the recipe too much. Similar things appear to be happening in some homes that are tightly sealed, walled off from outdoor life. In these places, we have changed the composition of life in ways that have broken sourdough's ecological system.

Our dysfunctional, laboratory sourdough starters are microcosms of our biological macrocosm. As for our own role? Earlier, I compared humanity to microbes on a Petri dish, but that isn't quite right, because we aren't alone on our round home. We are one species in the broader community of life, and yet we are a species with disproportionate effects. We humans are akin to the lactic acid bacteria in a sourdough starter. Like us, the lactic acid bacteria shape the world of which they are a part while also simultaneously being dependent on the other species around them. But unlike us, the lactic acid bacteria tend to make the world around them more hospitable for themselves. They produce acid and thrive in its context. There are also two more big differences. The first is that the lactic acid bacteria live in a world that contains tens of species, not millions, billions, or trillions. The second is that when the lactic acid bacteria use up all their resources, we save them. We reach down and offer them flour, anew.

IF WE RUN out of food, we will not be rescued by a celestial restocking of our stores. We must both use resources and sustain their production.

One might argue for a third difference between our role and that of lactic acid bacteria. We are self-aware, sometimes.

Our self-awareness, however, has its limits. Even once some of the consequences of our decisions begin to become apparent, our various actions are often so intertwined that it is hard to know which one caused any particular effect. Recently, a group of amateur entomologists in Germany began to revisit collections of insects they had made over the last thirty years. Those insects had been gathered using a standardized type of trap in standardized sites. Year by year, the insects from the traps were sorted, identified, and added to the group's collection. The initial goal of these amateurs, many of whom were, like Terry Erwin, beetle people, was to simply document the insects of Germany, with a focus on rare species. The amateurs did not necessarily expect to document any major surprises, and certainly nothing that would be newsworthy outside their small group. After all, Germany is one of the two or three best-studied places on Earth with regard to insects. Also, while it is more diverse than Linnæus's Sweden, it is not drastically so. There are almost certainly more insect species in an individual tropical forest in Panama or Costa Rica, for instance, than in all of Germany. For example, while there are about a hundred species of ants known from Germany, more than five hundred species of ants are now known from the forests found at La Selva Biological Station in Costa Rica.[13] Yet when the entomologists compared the number of insects they had collected in different years, they were in for a shocker. Over the preceding thirty years, the total biomass of insects in the natural habitats they were studying had declined by 70 to 80 percent, unnoticed. This had occured in one of the best-studied countries on Earth. It is still unclear just what caused this decline.[14]

It is also unclear what the consequences of this decline in the numbers of German insects have been. We know that it has led to losses of population in insect-eating birds. But what else has happened? No one yet knows. We will know, I suppose, the consequences once we run into them.

It is easy, with so much in flux, so much unknown, to give up. In the darkness of our ignorance and disorientation, maybe the easiest solution is to abandon ourselves to our fate, to walk blindly into the future, hoping. We can't figure it out. It is too complex, and we are too ignorant, and too much has changed. Sure, we are going to bang our heads trying to find our way, but maybe that is our lot. Another option is to focus on the details, to zoom in on the story of a particular German beetle species. From a deep knowledge of the specific, broader solutions can emerge. The focus on the specific needs to be part of the approach, but it will never offer a full picture, in no small part because there are just so many damned specifics.

The approach I embark on here is to avail ourselves of life's laws to make sense of our changing world even before we have named all of its parts. But even as we do, we need to keep Erwin's law in mind. Erwin's law reminds us that the biological world is bigger and more diverse than we imagine; the known world is modest and the unknown immense. Even the laws I will introduce to you in this book are subject to Erwin's law, subject to the possibility that the organisms that have not yet been studied don't necessarily behave like those that have been. And yet the knowledge that our view of the living world is fuzzy, partial, and biased shouldn't stop us from trying to use what we do know to make sense of the world. Amid the great darkness, our light is dim and yet nonetheless illuminating, and one way or another, we need to find our way.[15]

CHAPTER 2

Urban Galapagos

E. O. WILSON WOULD COME TO UNDERSTAND THE DETAILED WORKINGS of one of the living world's most robust laws, a law that predicts not only how fast species will go extinct and where, but also how fast and where new species will evolve, indeed, where they are evolving right now. But that isn't where his story starts. His story starts in Alabama where he grew up as a spindly, animal-loving boy. He loved snakes, sea creatures, birds, amphibians, and pretty much everything else that moved. One day, when fishing in Pensacola, Florida, he yanked his line too hard. A fish flew up out of the water and poked him in the eye, permanently damaging his vision. This incident left him unable to study and catch fast-moving vertebrates. Meanwhile, congenital hearing problems in the upper registers left him unable to hear the calls of many birds and frogs. He was, as he wrote in his autobiography, "destined to become an entomologist."[1] His focus—as a boy, as a university student, and eventually as a Harvard professor—was to be ants.

On one of his early journeys to study ants, Wilson traveled to the islands of Melanesia, including New Guinea, Vanuatu, Fiji, and New Caledonia. At the time, he'd been selected as a Junior Fellow in Harvard's Society of Fellows. This granted him freedom to research whatever he saw fit. So he went to Melanesia, where, in essence, he was paid to collect ants for science and to think. (I've had that job. It is a good job.) As he turned over logs, flipped leaves, and dug holes, through his one good eye he saw patterns in how many and which ants lived on different islands. Those patterns seemed to reflect rules of nature. Among the ants, Wilson felt as if he had grabbed on to electrifying and deep truths about the world. One of those truths was that bigger islands had more kinds of ants than did smaller islands.

Wilson was not the first to notice that bigger islands house more species. Other scientists had already detected that the distribution of bird and plant species followed such a pattern. This pattern could be depicted in a simple equation showing that the number of species on an island equals the area of the island, raised to some exponent, multiplied by a constant. In short, the bigger the island, the more species it can be expected to have. The ecologist Nick Gotelli calls this equation and the pattern it depicts "one of the few genuine 'laws' of ecology," the species-area law.[2]

It is often said that Sir Isaac Newton discovered gravity when an apple fell on his head. But this is wrong. Newton's great contribution was not the discovery of gravity but the discovery of the cause of gravity. E. O. Wilson was like Newton in that he was not satisfied to simply notice life's gravity-like pattern, the tendency of species to accumulate on big islands. He wanted to explain why and, in doing so, to develop ecology into a rigorous mathematical science of laws. But there was one problem. Wilson's math wasn't much better than his ability to spot snakes or hear birds, and so, as a Harvard professor, he took a first-year calculus class. Wilson knew he needed to learn, so he got to it, no matter that he had to scrunch his long legs up into a student desk, sit quietly, and work through his homework and quizzes. Also, knowing that freshman calculus alone wouldn't be quite enough, Wilson partnered with an ambitious young ecologist whose math was plenty good, Robert MacArthur. Together, MacArthur and Wilson began to develop a formal mathematical theory, a theory that the two thought might account for why there are more species on bigger islands, whether they be ants, birds, or anything else.

The theory had two key components. The first saw the probability of any particular species going extinct on an island as a function of the size of the island. MacArthur and Wilson thought that the chance that a species would go extinct from an island increased as the size of an island decreased. On smaller islands, populations of organisms were necessarily smaller, and so the chance that they would go extinct due to, say, one bad storm or one bad year was greater. What was more, the chance that a small island might not have enough of whatever the organisms needed was also greater. Time has lent support to the idea that there exists a general relationship between island area and extinction. The extinction rate of species on small islands tends to be higher than the extinction rate on larger islands, especially when those smaller islands have fewer kinds of habitats.

The theory's second component addressed not the loss of species from islands but, instead, their arrival. Species can colonize islands from elsewhere, whether by flying, floating, swimming, or catching a ride. Or they can evolve in situ. Wilson and MacArthur imagined that in both cases, the probability of such "arrivals" increases with the geographic area of islands. Species have a better chance of finding an island if it is bigger. Bigger islands are also more likely to have whatever special habitat, host, or other requirement a particular species needs. In addition, a bigger island might also provide more space for populations of a species to become sufficiently isolated from one another to evolve into different species.

MacArthur helped Wilson elaborate all of these ideas, expand them, and express them in a set of equations that they would publish in a booked called *The Theory of Island Biogeography*. Their theory would go on to be tested across the islands of the world. It was tested by tens and then hundreds of scientists, most of them graduate students excited to make sense of the world's hidden rules. The details of the equations have been contested, argued about, and fussed over, with the kind of fastidiousness scientists reserve for important things. MacArthur and Wilson's equations ignore many features of the biology of islands. Yet their theory has stood the test of time; it captures essential truths about the world's workings. Bigger islands do tend to have more species, and this does appear to be because of the balance of extinctions and arrivals. Perhaps just as importantly, their theory also offers clear predictions about the nature of the future, whether one is considering remote islands, wild forests, or even cities, especially cities.

IT DIDN'T TAKE ecologists long to realize that MacArthur and Wilson's theory should also apply to island-like bits and pieces of habitat, of which there are now many. How different, after all, is a patch of British woods in a sea of agriculture from a patch of rock and dirt in an actual sea?[3] And don't the medians in the middle of the road on Broadway in Manhattan form a sort of archipelago amid the sea of glass and cement? What is more, the extension of MacArthur and Wilson's ideas to patches of habitat seemed urgent. At the time, as now, forests and other wild habitats were being lost at alarming rates. And if MacArthur and Wilson's ideas about islands really applied to dwindling forests, many species were also, undoubtedly, being lost from them too. Was this story to be found in the fragments? MacArthur and Wilson predicted it was. This possibility sparked a series of massive research projects including a giant experiment, led by Tom Lovejoy, then at the Smithsonian Institution, to intentionally create fragments of forest in the Brazilian Amazon.

The author Terry Tempest Williams wrote, in considering our planet, "If the world is torn to pieces, I want to see what story I can find in the fragmentation."[4] This is what Lovejoy wanted, to learn from the fragmentation. Lovejoy's experiment created patches of forest by turning the landscape around those forests into pasture. The forest was going to be clear-cut anyway, tree by tree, by ranchers; Lovejoy convinced the Brazilian government and the ranchers to turn those cuts into an experiment. The Danish verb for "cut" is *skaere*, from the same root as the word for shards, *skår*. It was shards that Lovejoy would create with his clear-cuts, broken pieces of a fragile ecosystem that had once been whole. The shard-like patches in Lovejoy's project were to be of different sizes and at different distances both from each other and from the "mainland," which is to say, the larger contiguous forest. The results of these experiments are chronicled in David Quammen's beautiful book, *The Song of the Dodo*, as well as in Elizabeth Kolbert's *The Sixth Extinction*.[5] What Lovejoy and many collaborators ultimately found was that patches of habitat do act like islands in the sea. The smaller they are, the fewer species they contain. And as the forests and other wild habitats on Earth shrink, the number of species newly arriving in those habitats will decline, and the number of species going extinct will increase.

Though our understanding of the details and dynamics of the effects of habitat loss on biodiversity continues to gain nuance from ongoing studies, we already know enough to act.[6] Wilson and other conservation biologists have led a call to conserve half of Earth's terrestrial area as wild forest, grasslands, and other ecosystems. It is this half of Earth, Wilson argues, that is required to save the biodiversity we need now or will need in the future. He should know; he helped write the equation.

Most of the time, the dynamics of island biogeography are reliably predicted by taking into account just the arrival of species on islands or patches (colonization) and the loss of species from

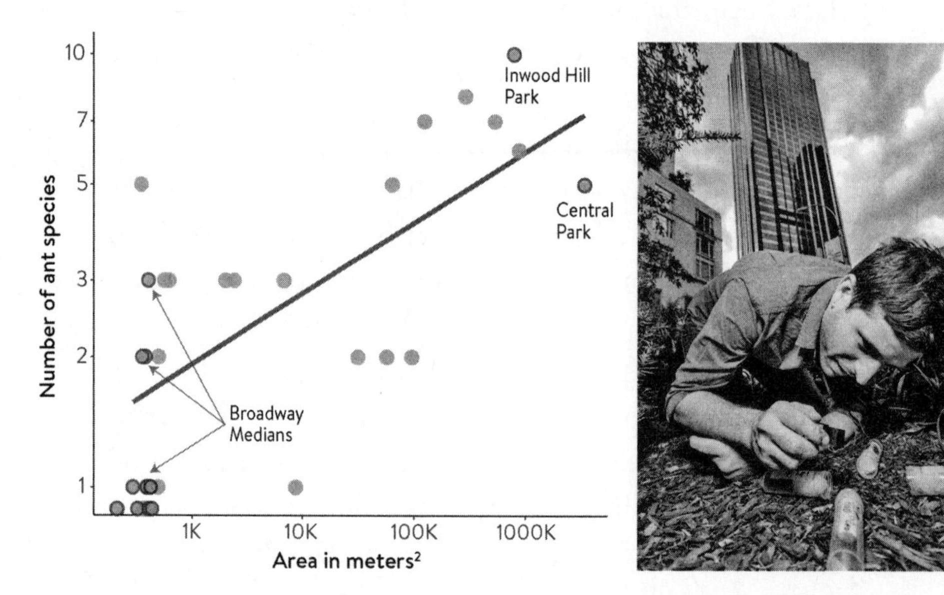

Figure 2.1. *Left*, one example of the relationship between species diversity and area for island-like habitats, ants in street medians and parks in Manhattan. *Right*, Clint Penick (at the time a postdoctoral researcher in my lab, now an assistant professor at Kennesaw State University) sampling ants in one of these medians by baiting them with sugar in small flasks. Figure by Lauren Nichols based on data from Savage, Amy M., Britné Hackett, Benoit Guénard, Elsa K. Youngsteadt, and Robert R. Dunn, "Fine-Scale Heterogeneity Across Manhattan's Urban Habitat Mosaic Is Associated with Variation in Ant Composition and Richness," *Insect Conservation and Diversity* 8, no. 3 (2015): 216–228. Photo by Lauren Nichols.

islands or patches (extinction). But there is another process in play, a process mentioned by MacArthur and Wilson but rarely commented on in subsequent studies, speciation.

Speciation is the formation of new species; it is the origin of two or more species where once there was just one. The rate of speciation is predicted to increase with habitat area. Wilson and MacArthur originally hypothesized that larger islands would have more arrivals, but they also predicted that speciation would be more likely and rapid on bigger islands. This prediction went little commented on in the years that followed the 1967 publication of *The Theory of Island Biogeography*. Maybe MacArthur and Wilson's thoughts on speciation were ignored because they were at the end of the book. But it is also possible that they were a little too far ahead of their time. Ecologists and evolutionary biologists did not yet realize how quickly evolution could occur, much less that the origin of species might be documented, as it has been, in real time.

If one does read to the end of the book, one finds that MacArthur and Wilson discuss speciation in some detail. They recognized that islands were "an excellent theater in which to study evolution,"[7] whether in the context of speciation, local adaptation, or even just the origin of novel traits. This sense of islands as evolutionary theater connected MacArthur and Wilson to Darwin. Darwin used islands as a lens through which to study evolution, but also as a context in which he could clarify his thinking. The isolated lands Darwin visited while on his nearly five-year journey on the ship HMS *Beagle*—including the Cape Verde Islands, the Falkland Islands, the Galapagos Islands, Tahiti, the islands of New Zealand, and the island continent of Australia—offered him a clear view of a circus of species he had seen nowhere else, species he would later realize had, in many cases, evolved on those islands. But islands also offered an ideal context in which to describe the workings of natural selection, rarefied stages on which to set his description of a process that occurs everywhere.

New species, Darwin argued, could evolve on islands in response to their isolation and to local conditions. The islands of the Galapagos archipelago, for example, were formed by volcanoes that rose up from the sea floor five hundred miles off the west coast of South America. A single species of medium-sized tortoise arrived on the islands and evolved into no fewer than fourteen species of giant tortoises, some bigger, some smaller, some darker, some lighter. A single species of mockingbird flew to the archipelago and evolved into three species, each on its own island. A single mutable and drab species of finch flew to the islands and evolved into thirteen species, now called Darwin's finches. The finches differed, as Darwin noted, in their beaks, which had become, as he would write in *The Voyage of the Beagle*, "modified for different ends" by natural selection.[8] One of the species of Darwin's finches evolved in such a way so as to use its beak to access nectar, pollen, and seeds from cactuses. Another became a vampire, pecking with its beak at the backs of birds and other vertebrates for blood. Another two species evolved the ability to use their beaks to hold sticks with which they hunt for grubs. Several species evolved beaks suited to a reliance on seeds.

Darwin thought that oceanic islands tended to be especially likely to have endemic species, species found nowhere else. Darwin realized that such species were present because, in their isolation, they had evolved differences from their mainland relatives. But Darwin was ambivalent about which kinds of islands might favor more new species and which fewer. MacArthur and Wilson added to Darwin's classical story of evolution on islands. What MacArthur and Wilson contributed was the hypothesis that the organisms that arrived on islands evolved into more species if the islands were bigger. But that hypothesis was hard to test. In fact, as of 2006 it was almost entirely untested, apart from one figure in MacArthur and Wilson's book, Figure 60. In Figure 60, MacArthur and Wilson plotted the number of species of birds on islands of different areas that were endemic to those islands, found there and nowhere else. There weren't many points on the figure, but the points did seem to suggest that there were more endemic bird species on bigger islands, perhaps because they were evolving there.

IN 2006, YAEL KISEL started her PhD studies at Imperial College working with Tim Barraclough, now a professor at Oxford University. Kisel would ultimately carry out the most ambitious synthetic study ever attempted of the effect of the area of an island on the probability that new species would evolve on that island. For millions of years, volcanic islands rose up out of the sea. Lava bubbled and then cooled. Algae colonized. Birds colonized. Spiders let out silk and allowed themselves to be carried away. They landed on the new land and colonized. Plants rode on birds' feet and floated on currents. And then evolution unfolded, shaped by the realities of the local circumstances but also by which species happened to have arrived. Kisel would study the aftermath.

Her study began as a side project. While Kisel was developing her main dissertation project, Barraclough suggested that she might also try to figure how small an island could be and still be big enough for one plant species to evolve into two over time. The effort would build on similar work that had recently been done on birds.[9] As Kisel put it to me in an email, she was trying to figure out if there was a "minimum island size for speciation in plants," and if so, what that size was. Eventually, Kisel and Barraclough decided to expand the project to other kinds of organisms, leading Kisel to gather more data until she found herself, almost accidentally, with the largest data set ever compiled of the characteristics of islands on which different groups of organisms had speciated. She'd compiled the entire thing without leaving Europe. She didn't go to the Galapagos Islands, Réunion island, or Madagascar. The work could all be done in museums and computer databases grounded in the fieldwork done by the people who had gone to those places.

Kisel's database contained data not only from small oceanic islands, such as the Galapagos, but also from much larger islands, the biggest being Madagascar. There were two kinds of speciation on which Kisel might have focused. She could have focused on whether a species, upon arriving on an island, evolved into a new species different from its relatives on whatever mainland it had come from. But that wasn't Kisel and Barraclough's main interest. They were more focused on speciation that had occurred within islands. By focusing on intra-island speciation, Kisel could consider not only the minimum island size necessary for speciation (her original question) but also what other factors might be important.

As she expected from MacArthur and Wilson's theory, Kisel found that the size of an island mattered to the likelihood of speciation. It was the single most important factor governing the probability of speciation in each of the groups of organisms Kisel studied. The bigger the island, the more likely it was that speciation occurred. But there was something else. Kisel developed a hypothesis, built on a mix of previous research and her own observations of the data: organisms that were poorer at traveling among or within islands should be more likely to be able to speciate on a small island. Conversely, organisms that disperse easily (and spread their genes everywhere) should very rarely or never speciate on small islands.

Kisel's logic was well reasoned. Organisms that disperse well, fly fast, run far, or even slither swiftly, might become isolated in different parts of a smaller island—for a while. But eventually, invariably, organisms from one part would meet up with those from the other. They'd mate, swap genes, and swamp any differences that might have been accumulating in one population relative to the other. Think of this in the context of dog breeds gone feral. Imagine that dogs of one breed, bulldogs, were released on one side of an island with an extreme habitat that might favor adaptation, and dogs of another, golden retrievers, were released into a more forgiving habitat on the other. As long as the island was small and the barriers few, some golden retrievers would invariably immigrate to the bulldog side of the island (and vice versa), breed, and produce offspring whose traits were a genetic blur of the two parents. Or, as Darwin put it, "any tendency to modification will also have been checked by intercrossing with the unmodified immigrants."[10] But if the island was big enough, the two dog populations might never meet. They might evolve over time, along separate trajectories, until they could no longer breed, so that even when they did find each other, they would remain separate. In short, Kisel predicted that for organisms that don't disperse well, even small islands would be big enough to allow speciation. But for good flyers, such as bats, or good walkers, such as mammals of the order Carnivora (including wolves and dogs), speciation would only be possible on big islands.

Kisel and Barraclough considered this dispersal hypothesis for the species of the different kinds of organisms in Kisel's big database—birds, snails, flowering plants, ferns, butterflies and moths, lizards, bats, and carnivorous mammals. Theirs was a hodgepodge of life-forms, but they were the life-forms for which data were readily available. Excluded were most kinds of mammals, most kinds of insects, and all of microscopic life. For all the life-forms Kisel and Barraclough studied, new species are more likely to evolve on bigger islands. But the minimum size of an island at which speciation is possible is smaller for organisms that disperse poorly (snails) and larger for organisms that disperse more readily (birds and bats). The minimum area for a snail to evolve a new species is tiny—less than a square kilometer, roughly the size of Tesla's factory in Fremont, California. Conversely, the area required for a bat, with the ability to fly far and wide, is many times bigger, some thousand square kilometers or more—roughly the size of New York City, including all five boroughs.

When her project on the evolution of new species on islands was done, Kisel moved on to other things, leaving a number of ideas untested. One of those ideas relates to the snails. Although they are seldom featured, well, anywhere, snails have evolved new species on islands around the world. Snails diversify especially readily. In part, this diversification might just be due to the slow dispersal of snails ("I think I can; I think I can"). But Kisel thought something else was also at play. As she explained to me in an email, to diversify on islands, species need two attributes. They need to be homebodies, so that they are able to avoid breeding with their relatives on other islands or on the mainland. But they also need to be able to get to the island in the first place. Snails have this combination in spades. Their average movements are very local and very slow; a snail might move no more than a meter during its lifetime. But every so often—at least, often enough to get to islands in the first place—they are carried long distances, on bird feet, in bird guts, and even on floating logs. Snails occupy the sweet spot with regard to the origin of species. Frogs, on the other hand, are likely to diverge if they get to islands, but they arrive only rarely. As Charles Darwin noticed, they aren't good at long-distance dispersal. Very few oceanic islands have native frog species.

The combination of occasional long-distance dispersal with average short-distance dispersal can occur in two steps, wherein a species that initially disperses well loses its ability to disperse once it arrives on an island. The loss of dispersal abilities is advantageous to species if it is, on average, better to stay on the island than to travel away from it, which is often the case. This is just what happened with the bats of New Zealand. A lineage of bats arrived in New Zealand but, having arrived in a forgiving environment surrounded by unforgiving seas, lost the ability to fly. Once flightless, the lineage was much more likely to diverge among habitats within New Zealand, and it did. Something similar happened with birds on many islands. Flightlessness has evolved in island bird lineages many times, and once it has, those birds then often diverged into multiple species. Such birds are rarely seen now, in part because once humans arrived on islands, flightless birds were especially at risk of being eaten, whether by humans or by the species that come with humans, including mice and rats.

KISEL AND BARRACLOUGH'S results and predictions allow us to reconsider what the theory of island biogeography has to tell us about the life around us. We should expect ancient species to be going extinct in shrinking patches of forests, grasslands, and swamps around the world. They are. In some of those patches, new species will also be evolving from populations now isolated from others of their kind. But the origin of such new species will be far rarer than the extinctions of existing species, both because the process of extinction is far faster than the process of speciation and because speciation is less likely in small patches of habitat than in big ones.

Meanwhile, species able to survive in the habitats that are now expanding should persist, riding with us into the future. For types of organisms that disperse well enough to arrive in our expanding anthropogenic habitats in the first place but not well enough to move among them, we might expect new species to have already evolved. Such types of organisms include, by Kisel and Barraclough's accounting, snails. But they also include some kinds of plants, particularly those whose seeds are not very good at dispersing—for example, plants that rely on ants to carry their fruits, such as trilliums, violets, and bloodroot. They also include many kinds of insects. As for even smaller life-forms, no treatise on their island biogeography has yet been written. Some fungi are very poor dispersers and have the potential to diverge among habitat islands, even small ones. On the other hand, some bacteria species disperse so readily in the wind that they are more like flying mammals, unlikely to diverge except when isolated by some unusual impediment. As for viruses, new strains, as we have seen recently with the virus that causes COVID-19, can evolve even just within an individual human body.

Kisel and Barraclough's work suggests the possibility of a bold new world evolving all around us, one in which the identity of the newest species is relatively predictable. Yet it is one thing to predict such a world, and quite another to show that it has come (or is coming) to be.

By far the largest habitats that humans have made are our farms. The collective area of corn planted on Earth is roughly the same size as the country of France; to corn-eating species, our cornfields are immense islands in an archipelago arrayed across continents

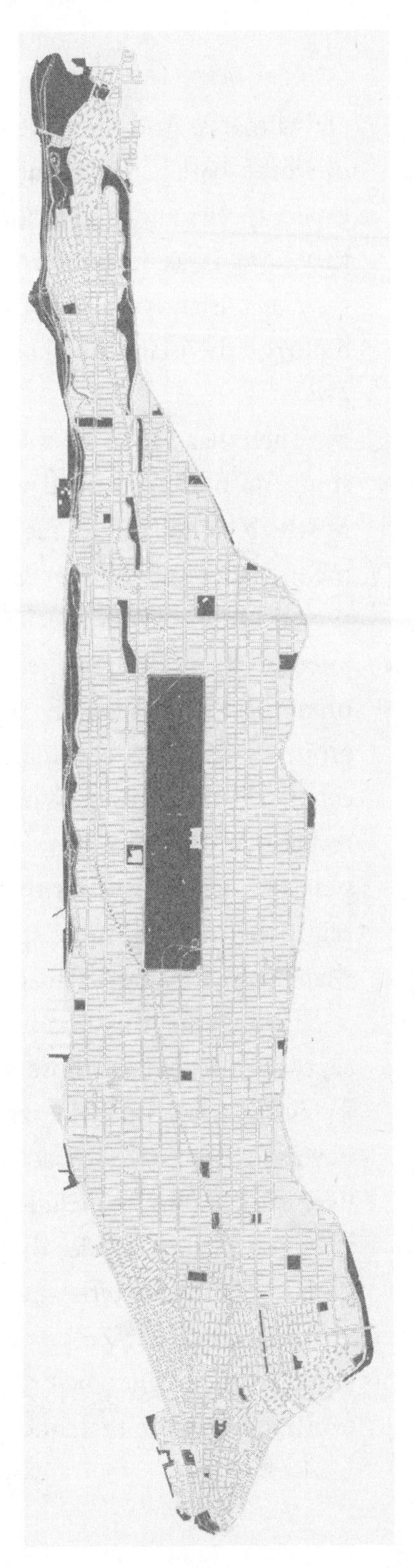

Figure 2.2. The archipelago of patches of green, such as medians and parks, against the backdrop of the island of Manhattan. For species that rely on grassland or forest habitat, these green spaces (shown here in gray) are island-like and, to varying degrees, isolated. However, for species that live in the less green spaces of the city, the world of streets, glass, and cement, Manhattan is one large and connected island filled with abandoned bits of delicious food. Figure designed by Lauren Nichols.

and climates. And there are other agricultural archipelagos too—of wheat, barley, rice, sugarcane, cotton, and tobacco. We would expect species endemic to these islands of crops to evolve. Indeed, they have. If, as the author David Quammen put it in *The Song of the Dodo*, islands are the "Dick and Jane primers of evolutionary biology," the island-like habitats provided by farms are *War and Peace*.[11]

There has yet to be a Charles Darwin or Yael Kisel of corn smut; no one has looked with wonder at our farms as contexts in which to holistically see the unfolding of evolution's wonderments. And frankly, that is a shame. What we do know about the evolution of new species in our crops comes from studies of attempts to understand such species in order to control them; often these studies are divided up by subdiscipline, such that one group of scientists considers the fungi, another the insects, another still the viruses. When considered together, these studies reveal that our crops now host hundreds, perhaps thousands, of pest and parasite species that live nowhere else. It is almost certainly the case that more species have evolved anew in our crops than evolved in the Galapagos Islands.

Throughout this book I use the word "parasite" in a broad sense, to include all those species that live on another species. Typically, when I use the term I am referring to species that also have some negative impact on the species on or in which they live. Such parasites include species of worms and protists, but they also include species that are often called pathogens, such as disease-causing bacteria and viruses. Some of the parasite species evolving amid our crops are their ancient associates that have clung to them since before they were domesticated. They then evolved, relative to their ancestors, changing as the crops changed.

As that happened, they became new species, distinct from both their ancestors and their living relatives.

Other parasite species, as well as pest species, colonized our crops anew, traveling to them from other habitats the way finches flew to the Galapagos. The ancestors of Colorado potato beetles lived on wild species of the plant genus *Solanum* in North America (the potato is native to South America). In the 1800s, the beetle colonized potatoes and, having done so, rapidly evolved tolerances for the climates in which potatoes are grown as well as resistance to the most common pesticides sprayed on potatoes. Colorado potato beetles now thrive basically anywhere there are potatoes in the Northern Hemisphere.[12] The species of *Phytophthora* parasite that caused the potato famine used to live on the wild species of the plant genus *Solanum* in South America but made the jump to domesticated potatoes, whereupon it evolved new traits and traveled to Ireland and around the world.[13] The parasite that causes wheat blast disease evolved from ancestors native to a Brazilian pasture grass, *Urochloa*. The grass was introduced to Brazil from Africa roughly sixty years ago, apparently with its parasite. Some individuals of the parasite made the jump from the grass to wheat; once on wheat, the descendants of those individuals evolved so as to be able to better take advantage of wheat. Their descendants, in turn, spread across Brazilian wheat fields, moving from plant to plant like a gust of wind.

Yet another type of origin of species in the context of agriculture can occur when breeders create new types of crops. In the 1960s, crop breeders produced a successful variety of a crop called triticale, a hybrid of wheat and rye. Very soon thereafter, that variety was afflicted by a new disease, a powdery mildew. The mildew is caused by the parasite *Blumeria graminis triticale*. The parasite is a new lineage. It evolved via hybridization between one parasite species that lived on wheat and one that lived on rye.[14]

Nor are all of the new species in agricultural fields just pests or parasites. New kinds of weeds have evolved that mimic crop seeds and are inadvertently sown by farmers, at least when seeds are harvested manually. New species have even evolved to take advantage of crops once they are in storage. The house sparrow (*Passer domesticus*) appears to have evolved from its wilder relatives into what we can realistically call a new species along with the origin of agriculture, around eleven thousand years ago. As it did, not only did it become separated from its wilder relatives, but it also evolved the ability to feed on a higher-starch diet, associated with our grains. Similarly, *Sitophilus* grain beetles evolved to rely on our stored grains. In doing so, they lost their wings. In addition, they evolved a special relationship with a new kind of bacteria species that came to live inside their guts, on which they rely for nutrition (specific vitamins) they can't find in grains.

The new pests, parasites, weeds, and other organisms that have evolved among our crops are not always referred to as new species. Sometimes they are called strains, varieties, or lineages. Typically, these are distinctions without a difference, subtleties of the agricultural subdisciplines charged with keeping track of who is eating or competing with our food. What is clear is that just as new finch varieties and then species evolved after having colonized the Galapagos and new bat species evolved after having colonized New Zealand, new varieties and species of pests and parasites are evolving all around us in the enormous islands of our farms. In each of these cases of colonization, adaptation, divergence, and the origin of species, the new species evolve genetic changes but also specific adaptive, physical manifestations of those changes. Darwin wrote about the "beak of the finch," but there is just as much sublime magic in the unfolding changes in the proboscis of the potato beetle or the secreted proteins of the mildew. As should be clear from these examples, the origin of new species in our fields is typically to our detriment. They eat, uninvited, from our plate.

IN ADDITION TO islands of agriculture, we have also created enormous islands of urbanization. Cities emerged so quickly relative to the ordinary pace of Earth's change that their growth was almost a sort of vulcanism, an eruption and solidification of cement, glass, and brick. To a great extent, evolutionary biologists ignored the evolution that might be happening in the midst of this tectonism. Remember that biologists have a tendency to pay the most attention to large mammals and birds. Large mammals, such as coyotes, move too well to become isolated in individual cities. Birds fly from city to city, or at least they sometimes can and do. But most species in cities are smaller and disperse less well. Smaller species often, because of their shorter generation times, evolve more quickly. And, as Kisel and Barraclough noted, species that disperse less well are more likely to become isolated and diverge. As evolutionary biologists have begun to pay more attention to cities, they've seen hints of divergence among the fast-evolving, slow-to-disperse species.

Rats are not the most likely group of organisms to evolve new species in cities. They have faster generation times and move less than do coyotes, but they are hardly snails. Yet recent research by my friend and collaborator Jason Munshi-South has shown that in some regions, geographically separate urban Norway rat populations are already diverging from one another, becoming ever more distinct—almost certainly as a function of the specifics of their cities, their climate, the available food, and other details.[15]

This is true not only for rats from very widely separated cities—such as rats from Wellington, New Zealand, relative to those from New York City—but also for rats in different cities in the same region. Munshi-South has recently shown that the Norway rat populations of New York City are closely related and show almost no evidence of breeding with Norway rats in nearby cities. More than that, the rats at one end of Manhattan appear to be diverging from those at the other end. Norway rats are less likely to travel through, eat in, mate in, or live in Midtown Manhattan, perhaps because Midtown has a lower density of permanent human residents than the other parts of Manhattan and hence also less of the rat food such inhabitants graciously, if inadvertently, provide. Whatever the reason, Midtown is, from a rat's perspective, a kind of sea between two lovelier islands. Similarly, Norway rats in one part of New Orleans are isolated from those in another by waterways and are diverging as a result. Norway rats in parts of Vancouver, on the other hand, have become separated from those of other parts of Vancouver because of difficult-to-cross roads. If current patterns of mating and movement continue, each city will eventually have its own unique Norway rat species, adapted to the circumstances of its local surroundings, part of the outdoor terroir of each city.[16]

House mice, after having spread around the world with humans, have now diverged into a number of new species and many more varieties. So far, these species and varieties differ only in details. None is yet radically different from the others, but give them time. This phenomenon of diversification among cities has been less well studied with houseflies, yet it looks as though houseflies in different regions of North America are adapting to their local conditions. My prediction is that diversification is also occurring in many smaller species but has just gone unstudied. We are blind to the changes happening around us.

The more different cities become from the habitats that surround them, the more they will act like islands. This applies to the evolution not only of new species but also of those species' traits. As I've already noted, one trait common among island species is loss of the ability to readily disperse, such as birds' loss of the ability to fly. On a remote island, a bird or a seed that travels too far from home is more likely to find itself in the ocean than in a good habitat. We might expect species on urban islands to also lose their abilities for transit, at least where nearby conditions are predictably more favorable to them than faraway ones. Some urban populations of the holy hawksbeard plant have already evolved the tendency to invest less in seed dispersal than do their more rural relatives.[17] They stay nearer to home. Species that lose their ability to disperse among patches of habitats are even more likely to diverge into new species, different from one city, farm, or waste-treatment plant to the next.

In the future, the ways in which borders are controlled will shape the fate of the many species evolving in cities. When we become better at controlling the movement of species around the world than we currently are, the species in our cities will diverge from one another even more readily. This could happen if we were to implement border controls. It could also happen if the global economy were to collapse, meaning that fewer people would travel from place to place. It is happening right now, to some extent, due to the virus that causes COVID-19. In any of these cases, the evolution of species would probably come to match our political regions, or at least the regions across which we enforce control. The species of European farms or cities, thus, might come to be different from those of North America. As far as I know, no one has yet looked, but it seems very likely that such differences are already accumulating in countries, such as New Zealand, that work hard to prevent unwanted species from crossing their borders. Such differences are also likely to emerge on either side of borders that are sealed by war or political strife. North Korea may well have unique agricultural and urban species that have evolved since the end of the Korean War.

New species might also form within cities by specializing in particular habitats. This is actually the more direct analogy to the cases that Kisel and Barraclough considered. It is also akin to what happened in the Galapagos Islands when a lineage of terrestrial iguanas evolved the ability to take advantage of the life beneath the surface of the water. The iguanas evolved shorter legs, flatter tails, and other adaptations that better enabled them to dive to the bottom of the sea to reach algae that few other animals ate. They evolved new kinds of spines and a lavalike gray-black skin that Darwin noted led them to be described as the "imps of darkness." Similar divergence, of even more impish life-forms, is happening now in cities. In Africa, urban populations of two species of *Anopheles* malaria mosquitoes appear to be diverging from their more rural counterparts, potentially because the urban forms have to evolve tolerance for the pollutants so prevalent in human cities. In London, populations of *Culex pipiens*, a species of mosquito, moved into the London Underground railway system (the Underground) in the 1860s. Since then, these mosquitoes have diverged so much from their aboveground relatives that they are now regarded by some as a separate species, *Culex molestus*. The aboveground species is adapted to feeding on birds. The belowground species is adapted to feeding on mammals (humans, rats, and the like). Females of the aboveground species require blood in order to lay eggs; females of the belowground species, where food is scarcer, do not.[18]

The world indoors has even more potential for being an epicenter for the origin of new species. My collaborators and I have found roughly two hundred thousand species in homes. Not all of those species only live indoors, but many do. Considering just animals, they include house centipedes, several dozen species of spiders, German cockroaches, and bedbugs. I estimate that no fewer than a thousand animal species now live primarily indoors. Many of these species are diverging between and within cities. This is almost certainly the case, for example, for house centipedes, which are now found virtually everywhere on Earth but don't seem likely to move very often. What about the most common house spider species? What about introduced species of ants that live primarily indoors, such as the ant *Tapinoma melanocephalum*? No one has studied the evolution of any of these species.

Then there are the life-forms most immediate to us, the species on and in our bodies, as well as on and in the bodies of the animals on which we depend, whether those be cats, dogs, pigs, cows, goats, or sheep. Many of the species that live on our bodies evolved as human populations grew. During the great acceleration of human population growth, the growth of populations of domestic animals also accelerated. While this was occurring, ever more of the species that sometimes relied on humans or domestic animals became more specialized. For such species, we and our animals were the meal ticket into the future. As ancient humans spread around the world, the species living on them diverged into new subspecies and in some cases new species. In studies that my friend Michelle Trautwein, a curator at the California Academy of Sciences, and I have done, we have found that as humans moved around the world, their face mites diverged.[19] Something similar has happened with lice, tapeworms, and even the bacteria on human skin and in human guts.

Of course, what I've just described is the scenario that is unfolding around us, not necessarily the scenario we desire. In many ways, a consideration of island biogeography leads to the conclusion that we have so squeezed, parted, and re-formed the wet dough of Earth that we have inadvertently extinguished wild species on which we depend or might depend and, at the same time, have favored the origin of species likely to cause us problems. And because extinction proceeds manyfold more quickly than the origin of new species, the numbers aren't even. We were offered a deal by nature: if we gave up thousands of species of birds, plants, mammals, butterflies, and bees, in exchange we could have a handful of new kinds of mosquitoes and rats. It is a bad deal, but one that so far we have accepted.

The good news is that it isn't too late to conserve more big, wild parts of Earth. Even saving half of all Earth, as E. O. Wilson proposed, is not out of reach. Such conservation can happen in parks, but also in backyards. Our lawns favor lawn-loving species. Get rid of your lawn and plant and favor native species; make your lawn part of the island of the archipelago that sustains native, forest, or grassland species. The bad news is that the threat to species from habitat isolation does not act alone. While we were cutting down forests and paving swamps, we also began to warm the world.[20]

CHAPTER 3

The Inadvertent Ark

AS CLIMATES CHANGE, SPECIES IN PATCHES OF HABITAT, HOWEVER BIG or small, will be faced with relatively few choices in how to deal with that change. Some species can deal with novel climates by modifying their behavior. Some diurnal species, for example, will begin to live by night. Other species can evolve a tolerance for new conditions. Most species will need to move. Let me say it again for emphasis: most species on Earth will need to move to survive climate change. Several thousand mammal species. Many thousands of bird species. Hundreds of thousands of plant species. Millions of insect species. Some untold number of microbe species. They will need to move from their present islands of habitat to different islands of habitat where their preferred conditions can newly be found. They will need to move to find their new home, a behavior that biologist Bernd Heinrich has recently described as homing.

Homing will become one of the most ecologically important phenomena of the coming centuries and even millennia. As tropical climates warm, tropical species will need to move to higher, cooler places, where they will also face more competition because the area of land decreases as one moves uphill. Or they can move north, if they are in the Northern Hemisphere, or south, if they are in the Southern. The species of Costa Rica, for example, will have to move toward parts of Mexico. Species of Mexico and Florida, meanwhile, must move toward, say, Los Angeles and Washington, DC. Even for species that can fly, this homing isn't easy.

Species need to figure out where their new homes will be, and then they need to get there. And unless they happen to be good long-distance flyers, they must move bit by bit, walking or riding from one patch of habitat to the next until they get to the conditions they need, if such conditions still exist at all. For many species, there will be no new home. They will wander and never find what they need. Or they won't find it in time. Or they will arrive and the climate will be perfect, but something else will be missing. They might arrive but be alone, without a mate.

A number of years ago, some of my colleagues at North Carolina State University and I decided to try to figure out the routes by which species might make these moves. We wanted to trace the routes along which they might travel. I would, for reasons that will become obvious, come to think of this effort as the Charlanta project.

The perspective of the Charlanta project team was informed by two ideas: the niche and the corridor. The ecological niche is a concept that was developed by the ecologist Joseph Grinnell in the early 1900s, by analogy with the small spaces in buildings designed to hold statues. To Grinnell the ecological niche was the small space in nature that held each species.[1] That each species has a niche is a law of life.

A niche for a statue only needs to be big enough and more or less the right shape to hold the statue. The niche that holds a species, in contrast, needs to accommodate all of its needs, whether with regard to food, climate, or places to sleep. As we think about the future, the most important of these needs has to do with climate. Each and every species has a set of climatic conditions in which it can survive. Some species have narrow climate niches; others have broad climate niches. Mountain lions, for example, have a wide climate niche: they can live in hot, wet rain forests, deserts, and cold temperate forests. Polar bears or emperor penguins, by contrast, have very narrow climate niches.

In light of climate change, ecologists have hurried to characterize, one by one, the climate niches of many species. In doing so, they learned a trick: measuring the climates in which a species lives today gives you a pretty good predictor of its climate niche. In addition, if you know this measure of the climate niche of a species, it is possible to predict where that species will be able to survive in the future, as the climate changes. It is possible to predict where, in homing, it must go.

The second idea that informed our thinking is the concept of the conservation corridor, a kind of natural habitat bridge that a species uses to get from one place to another, be it one city park to another or one continent to another. Corridors are constructed by conserving the habitats species need. When used to help species move, corridors are tools. But corridors can also be used as a rule to understand which species will succeed in the future. When the use of corridors as a tool for conservation was first proposed, contentious debate ensued.

My friend Nick Haddad was one of the early proponents of the value of corridors to conservation. Nick is a conservation biologist whose work has come to focus on the conservation of rare butterflies. When he was still a graduate student, Nick began to argue that corridors might both serve as habitats in and of themselves and, at the same time, help species, including butterflies, move from point A to point B. Nick could close his eyes hard and envision kaleidoscopes of butterflies and herds of mammals moving along corridors of forest or grasslands. Flightless mammals and insects would walk. Small birds would fly. Seeds would ride on and in the mammals and birds. And insects too, the multitudinous insects, would make their way. In the context of climate change, this parade of life would invariably proceed from a point nearer to the equator to a point farther from the equator, or a point at the bottom of a mountain to one farther uphill. Logical enough, at least to Nick.

The concept was initially met with criticisms that were both reasonable and difficult to test. Some argued that corridors would tend to be too narrow, all edge and no middle, and hence full of species from adjacent habitats. Others argued that species wouldn't use corridors or that corridors would favor the movement of invasive species but not native species, or of animals but not plants. The more time scientists had to find flaws in the potential of corridors, the more flaws they found.

The trick was in finding a way to test whether corridors worked. Nick Haddad had an idea. Nick likes working outdoors in the field. And he likes building and tinkering—whether that means replacing the pipes in his old house or building some device he needs to carry out his work. Nick is often carrying a hammer or a wrench. Nick, the builder, had an idea about a way to construct corridors. He wrote a proposal to the National Science

Foundation to travel to a site in South Carolina where the US Forest Service was regularly logging trees, the Savannah River Site. At that site, Nick would work with the Forest Service to saw out "islands" of grassland habitat by cutting down trees to reshape habitats; it wasn't quite carpentry, but nearly so. Often, when people think of island-like patches of habitat, they think of forests in fields or forests in grasslands; these would be the opposite, island-like patches of grass in a sea of forest. Such patches are common in nature. Picture a grassland after a small forest fire, a grassland surrounded by forest. Picture a meadow that has grown over an old, dry pond. Picture a patch of open ground, at the top of a hill, just above the trees. Nick would connect half of his island-like patches of grassland via corridors, which he created by cutting down more trees. The result looked like a cartoon barbell. Meanwhile, other patches would stay disconnected. In essence, the goal was to create two sets of replicated worlds, one connected by corridors, the other not. (There were other complexities that Nick proposed to consider too, but they were subplots on the main theme.)

The reviewers of Nick's grant said it was impossible, especially for someone so young, to do the work that had been proposed, more a "dream" than a plan. The grant was rejected. However, Nick was able to find other ways to fund the project. The project was not, Nick would prove, impossible. Instead, it became the most important experiment on corridors ever conducted, an experiment that continues to this day.

Nick created the patches of habitats and their corridors, and then he began to study how and whether species moved through them. Initially, Nick did the work with his wife Kathryn Haddad. The Forest Service created the habitat patches and corridors and, nets in hand, Nick and Kathryn documented their ecology, focusing on butterflies. After a while, to Kathryn's relief, Nick was able to find research funds to hire a team. Dozens and ultimately more than a hundred scientists would work with Nick to study the corridors. They studied butterflies, birds, ants, plants, rodents, and much more. What they found was good news. With certain caveats, the corridors worked. In dozens of scientific papers, written with his students, collaborators, and as time cemented relationships, friends, Nick wrote about the details of these workings.

While Nick was studying these corridors, other scientists had begun to study how animals moved through corridors on an even larger scale. They were seeing evidence that if you left or created enormous corridors for jaguars, jaguars would move through them (which is just what happened when jaguars returned to the southwestern US). Native wild mice move through cities only along the narrow corridors of green life—path to park to urban lot—that wind through them.[2] Eventually, even many of Nick's early critics would come to begrudgingly note the benefits of corridors, especially for relatively mobile species such as butterflies, mammals, and small birds. In part, their support of the approach was due to the results of Nick's experiment and others like it. But their support also reflected changes in the conservation problems at hand. When Nick began his work, conservation biologists were most worried about conserving species in particular places and about how to connect patches of habitat in those places. But, in the last decade, in light of the increasing awareness of just how many species will have to move with climate change, the focus has come to be on getting species from where they are to where they need to go, not just on sustaining their populations where they are.

Corridors are now viewed as one of the most important tools we have for helping ensure that species can move in light of climate change. And conservation corridors are being added to around the world, often at large scales. The Y2Y corridor project, for instance, aims to increase the connection of wild habitats from Yellowstone National Park all the way to the Yukon Territory in Canada. And corridors, whether continental or more local, bring a set of additional benefits as well. Crops along corridors are more likely to be pollinated by native bees flying along the corridors. Pests are more likely to be controlled by the predators and parasites hanging out in corridors. Rivers fringed by corridors of trees have better water quality. And, in addition, corridors provide ways for people to move through diverse natural habitats. The forest along the Appalachian Trail, for instance, is both a corridor for wild species and a route for human exploration. Corridors aren't the only way to move species. Some species, for example, will be moved individually, helicoptered or driven from their old homes to their new ones. But once one realizes that millions of species need to be moved, corridors are among the only approaches that are practical.

Corridors are invariably compared to arks. In the ancient Mesopotamian story of the ark, later retold in the Bible and the Koran, a man is instructed to build a very large, round boat out of rope and pitch and to put members of each species on that boat to save himself and nonhuman life from the floods. In some of the earliest versions of the story, the floods are the work of a regretful god who has been bothered one too many times by humans. People are to be punished for being too loud, too numerous, and too annoying. The floods come. Horror ensues. After the waters retreat, Earth is repopulated, its wild biodiversity restored, by the descendants of the species that have ridden from one time to another, from before to after, and from one place to another, on the ark.[3]

If corridors are arks, ships to ferry life from here to there, from before to after (whatever "after" might entail), Nick's role is obvious. He is the ark's carpenter. This analogy pleases Nick; he is glad to be doing the good work of moving species, especially the butterflies that have long been his focus. He is quick to note that he has done none of this work alone, that the work has required tens, perhaps hundreds, of other carpenters. As for the species able to get on board, the Mesopotamian story of the ark, like the later biblical story, ignored the insects. Nick will make no such mistake. Meanwhile, while Nick was busy building one kind of ark, our collective daily actions were very rapidly building another.

IT IS OFTEN argued that our modern lifestyles make it harder for species to home in on the new niches they must find to survive climate change, that we have, in fragmenting the world, ruined the corridors through which species might move. But that isn't quite right. The truth is that our daily actions are destroying corridors, but they are also creating corridors, yielding a sort of inadvertent ark. While conservation biologists were busy trying to connect forests to forests, grasslands to grasslands, and deserts to deserts, the rest of us were connecting cities to cities. This became clear to me during the project we undertook to identify the routes along which species must move in the southeastern United States as part of the Charlanta project.

That I worked on the Charlanta project at all must, on some level, have been due to Nick's influence. At the time, Nick's office was two doors down from mine. He was close enough that when he laughed or talked loudly, I felt it through my wall. As a result, I heard the word "corridor" every day I went to work. Nick worked on corridors. His students worked on corridors. We talked about corridors in the corridor. Whatever the origin of the Charlanta project, our goal was to consider how cities would grow in the future and then to examine where corridors of natural spaces might persist. The effort was led by Adam Terando, whose office was, at the time, one hall (corridor) away from Nick's and mine. Curtis Belyea, whose office was next to Adam's, made the maps. Jen Costanza helped think about the wild habitats. My colleagues Jaime Collazo, Alexa McKerrow, and I played more supporting roles.

The standard way to predict the future of urbanization, climate change, or any other change in which human behavior plays a role is to consider different scenarios. "What if," the scientist asks, "we imagine a scenario in which humans do this or that or the other?" After imagining a series of "what if" scenarios, scientists then go about predicting the consequences of those scenarios, whether for wild species, cities, or climate.

In the case of our study, our "what if" was, "What if people keep doing what they have done in the past?" This was a "business as usual" scenario, the least imaginative possible vision of the future and yet, also, undeniably, the most likely. We modeled what would happen if the rules for where people could build houses were left unchanged, if people continued to prefer the same habitats as in the past (forests versus grasslands, hilltops versus valleys), and if roads were extended to connect new growth following time-tested patterns. Our model predicted that Charlotte and Atlanta will grow by about 139 percent in size and merge with each other and with other cities to form a single megacity, Charlanta, stretching from Georgia into Virginia.[4]

This growth is predicted to have a variety of effects on the connectedness of habitats and hence on what bits and pieces of corridors will remain for wild species. Forests of all types will become less connected, as will grasslands. There will be fewer good, long corridors of either habitat type. The effect on wetlands will be more modest, in part because current policy makes it more difficult to build in wetlands and that policy was built into the model. The big-picture result of the work was that if cities continue to grow as they have in the past, it will be considerably harder in the future for species to make their way through forest and grassland. Indeed, in the years since we made this model in 2014, it has already become harder. The good news is that during that time people have worked hard to buy and conserve the land necessary to connect this landscape, to give species a way to get where they need to go. The bad news became apparent as we stared at Curtis Belyea's maps.

Adam, Curtis, Jennifer, Alexa, Jaime, and I looked at the maps Curtis made and saw the natural places. The areas that were not natural were the proverbial "white space," the space that framed and interrupted the habitats on which we were focused.

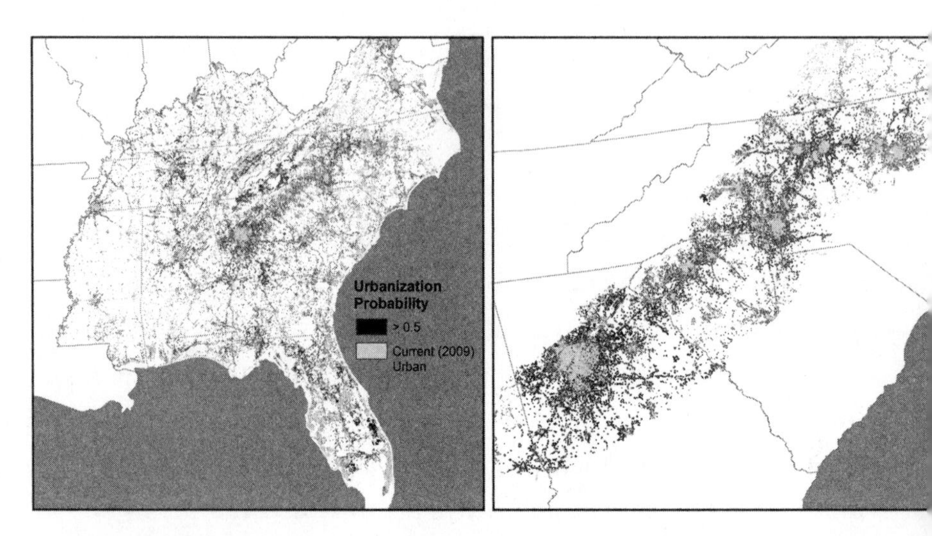

Figure 3.1. At left, urbanization in the southeastern United States as measured in 2009 (gray) and as predicted for 2060 (black). The panel at right zooms in on the future city of Charlanta, crawling over the Southeast like some giant anthropogenic caterpillar. Map produced by Curtis Belyea.

This was what most ecologists would probably have done; it is the time-honored bias of people in our field. Ecologists my age and older were trained to look to wild nature. In part, as the historian of science Sharon Kingsland has argued, this focus on the wild was a conscious choice by the founders of the field of ecology.[5] The founders chose to avoid the messiness of the everyday ecology of cities and farms, the messiness of the human-centered world. But that isn't the whole story. The focus also has to do with the people who choose to be ecologists. Many of us were kids who, like E. O. Wilson, grew up catching snakes and wading in swamps. Many of us are happiest when far from people. It isn't that we are misanthropic (though you can find a bit of that too) but, instead, that we have a fondness for big trees, unpredictable mammals, and narrow paths. When ecologists retire, they don't go on cruises. They move to cabins, where they keep doing their research while also often taking up a hobby such as raising longhorn cattle, drawing maps of forgotten places, carving chainsaw art, or assembling the world's largest collection of rare pomegranate varieties (to take a few examples from among my retired friends). This natural tendency has its benefits, but it also has its costs. One of the costs is that ecologists can sometimes miss things in plain view; they can miss the city for the trees. In Chapter 2, we have already seen that this has been the case for the story of islands and island-like habitats. As my collaborators and I looked at the maps Belyea made, we realized that it is also the case when it comes to thinking about how species will respond to climate change. To a great extent, we have already determined which species will be able to respond to climate change by moving. Our business-as-usual actions have built an ark, one likely to carry a specific set of species from here to there and from before to after. The ark is Charlanta.

If you look at Figure 3.1, you can see the nature of this ark. The right panel shows Charlanta, a megacity that will connect existing cities like knots on a string. But at its northern end, it also almost forms a connection with the megacity that already exists, the urban space that stretches from Washington, DC, to New York City and nearly, though not yet completely, to Boston. This is what we had missed. We have already created a corridor, a perfect and immense corridor, but it is not a corridor for rare butterflies, jaguars, and plants. It is, instead, a corridor for urban species, species able to move along roads and live amid buildings, species that live not in green spaces but in gray ones. As a result, the species that will be able to move to find their new homes will be species that thrive in cities, fly well, walk fast, or are predisposed to catching a ride not in the gut of a black bear or on the leg of a carrion beetle but with us, whether on our bodies, on the bodies of our domestic animals, in our vehicles, or even in our goods.

In the most ancient stories of the ark, a bird, often a dove, flies from the ark and does not return, having found and stayed on land that has emerged after the flood. The missing dove symbolized the time after the flood. Doves also offer a message about our future, thanks to research carried out by Elizabeth Carlen, a PhD student at Fordham University, and her adviser, Jason Munshi-South. In North America, rock doves, aka pigeons, thrive in urban landscapes but don't do terribly well in forests and grasslands. In eastern North America, the cities where they live are mostly connected by the Washington to New York urban corridor. However, there is a modest break in that corridor between New York and Boston. Recently, Carlen studied the genetics of the pigeons of North American cities. She found evidence that the pigeons from Washington to New York City are interbreeding so freely that there is no difference between DC pigeons and Broadway pigeons. Dispersal from one place to the other happens readily and easily. However, the pigeons from the DC to New York corridor are genetically differentiated, slightly, from those of Boston. They lack a sufficient corridor, for now.[6]

In the example of the pigeons of Boston, we can see the ways in which cities both allow organisms to travel and allow for the evolution of new species. Pulling together the concepts of island biogeography and corridors, we can predict that cities that are well connected to each other as megacities will allow species to move south to north (in the Northern Hemisphere). But the species in any particular megacity might be expected to diverge from those in other megacities. Meanwhile, the extent to which the story of any particular species is a story of dispersal, divergence, or extinction will depend on its population size, how readily it moves, and whether it has arrived in a particular habitat in the first place.

Our urban corridors are perfectly suited to ensure the survival of species that prefer urban habitats and disperse well. It is for them that we have inadvertently built an ark. But not just them. We have also connected the habitats of our homes and even our bodies. We have created corridors through which the bedbugs of the world can make their way north or south to their preferred climates. German cockroaches have a narrow climate niche; in China, they can only live indoors in buildings with air-conditioning and heat. A recent study argues that these cockroaches spread across China over the last fifty years via the corridors provided by climate-controlled trains.[7] Pigeons, bedbugs, cockroaches—not only have we connected these species and their habitats but we are already laying the groundwork for the future of their connectedness. We are investing in infrastructure that ensures their survival.

Depending on when you are reading this, it all might feel somewhat familiar. After all, we are currently in the midst of a very concrete manifestation of the fact that we connect regions of Earth not only by road but also by plane and boat. Globally, our coastal cities are connected by an extraordinary number of ships and shipping routes. Our cities are connected by an even larger number of flights. We have woven countries together through transit. And in the process, we have made another kind of corridor, one for a narrower set of species: those that can ride on and in human bodies. The coronavirus that causes COVID-19 traveled across these corridors, its path outlining the movements of human bodies from there to here and here to there again. This connectedness has great consequences because, as I will consider in Chapter 4, one of the reasons for the global success of humans has been our ability to flee from and escape the species that like to live with us at our expense.[8]

CHAPTER 4

The Last Escape

AS ANIMALS MOVE, SEARCHING FOR THE CONDITIONS THEY NEED, they will encounter species with which they have never interacted before. Species that have never been together will meet each other. Plants will meet new pollinators, but also new pests. Owls will hear other owl species they have never heard before. Mice will meet new mice. Each such meeting is an opportunity for a new story to unfold; there will be millions of new stories. Some of these stories are unpredictable, part of an unscripted drama unfolding all around us. Others, though, can be anticipated. Some of those that can be anticipated relate to the law of escape.

The law of escape states that species benefit when they escape their predators and parasites, their enemies. Species have long experienced the benefits of escape when they have moved to regions where their enemies are absent, have evolved so as to be resistant to their enemies, or, in rarer cases, have extinguished their enemies. In the last hundred years, such escape has been particularly conspicuous with regard to species introduced by humans from one region to another. Introduced species often flourish in the absence of their enemies. For example, many introduced trees are less eaten by herbivores than are native trees.[1] They are greener thanks to the enemies they have left behind. Humans are no exception to the law of escape. As we have moved around the world, we have benefited from release from our enemies.

Sometimes our escape has been from predators. Our ancestors were long besieged by predators. To the extent that wild non-human primates utter anything like phrases, those phrases tend to be along the lines of "Ooooh, delicious fruit"—a common chimpanzee exclamation—or, in species such as vervet monkeys, "Oh shit, leopard," "Oh shit, snake," and "For the love of god, giant baby-eating eagle!"[2] Early hominins were also eaten by leopards, snakes, and eagles, to name just a few of their assailants. One of the best-preserved early hominin skulls is that of the Taung Child; the skull is remarkable in that it appears to have been found beneath a giant eagle's nest, and it has talon marks in one of its eye sockets. Elsewhere, a number of hominin skeletons have been found in what were initially thought to be shelters but were later revealed to be giant hyena bone piles. Our ancestors, in short, were often eaten. Our modern fight-or-flight responses evolved in the context of this drama. But once our ancestors began to hunt, they began to kill off their predators.

As a recent study by the herpetologist Harry Greene and his collaborator Thomas Headland points out, some human populations are still subject to predation by giant snakes, but these are extraordinary exceptions.[3] For the most part, our escape from predators is complete, a dramatic story about the past. The same cannot be said of our escape from parasites. Part of our escape from some parasites is due to vaccination, hand washing, water-treatment systems, and other public health measures. But alongside these relatively recent escapes, humans also benefit, or fail to benefit, from escapes of a more ancient kind, escapes due to the geographic region in which they happen to live. As the world warms and species move via the connections we make among regions and continents, the benefits some humans experience from escape will become apparent, but only as those benefits disappear.

When one considers the entire world, the geography of our escape is relatively simple. As my friend Mike Gavin and I, along with two other collaborators, Nyeema Harris and Jonathan Davies, were able to show a number of years ago, infectious diseases of humans and the parasites that cause them are and have long been most diverse where conditions are hot and wet.[4] In this, they are not unique. Nearly every group of organisms so far studied is most diverse in the tropics, where conditions are hot and wet. Such conditions favor the diversification and persistence of species of beautiful birds, strange frogs, and leggy insects but also the deadly parasites that cause disease, including viruses, bacteria, protists, and even a small circus of monster-headed worms. Drier conditions, even if they are hot, are not as suitable to most parasites. Nor are colder conditions. Even when parasites that evolved in the tropics are able to survive in drier or colder realms, most are less likely to thrive. Simply put, the warmer and wetter a place is, the more kinds of parasites one encounters there and the less of an escape from parasites humans tend to experience.

As one zooms in on a particular parasite species, however, things get more complex. In many ways, malaria is emblematic both of the ancient geography of parasites and of such complexity. Today, malaria kills around a million people every year, but not everywhere, not in seasonally cold or dry regions where it is easier to control. It is a tropical parasite that some humans escape by living outside the tropics. This geography of infection and escape has ancient and tangling roots.

EACH MODERN SPECIES of African hominid, including gorillas, chimpanzees, and bonobos, hosts its own species of malaria parasite. As hominids evolved and diverged from one another, so too their malaria parasites. The malaria species that plagued the earliest human species (such as *Homo habilis*) probably included an ancient malaria species most closely related to the malaria species infecting modern chimpanzees and bonobos (much as we are most closely related to chimpanzees and bonobos). This was our ancestral human malaria, an heirloom of our origins. However, roughly two to three million years ago, an ancient human species appears to have evolved a change in the gene that produces a type of sugar found on red blood cells, the sugar to which one type of malaria parasite binds; that change made them immune to this ancient malaria for millions of years.[5]

Roughly ten thousand years ago, somewhere in tropical Africa, a strain of gorilla malaria made the jump to humans. In doing so, it evolved the ability to cope with human red blood cells and their lack of key sugars.[6] It also diverged, eventually becoming a new species now called *Plasmodium falciparum*, or just falciparum malaria. Falciparum malaria spread across Africa and kept going, its spread facilitated by the rise of agriculture with it settled populations of human hosts, often alongside standing water. It is falciparum malaria that now accounts for the vast majority of malaria deaths in the world.

The entire story of the evolution of the ancestral human malaria and humans' evolutionary escape from it occurred within the tropics. Similarly, the story of the colonization of humans by malaria parasites from gorillas and the evolution of falciparum malaria also took place in the tropics. As long as humans lived in hot, wet climates, they were potentially susceptible to malaria parasites and their evolutionary dramas, in which human bodies were the stage for a tragedy that was acted out again and again. Falciparum malaria has been sufficiently deadly over the last ten thousand years that some human populations have evolved adaptations that make them less susceptible to the parasite and its consequences.

When humans moved to drier or cooler habitats, they moved themselves off the stage of malaria's persistent tragedy. Both the malaria parasite itself and the mosquito species that carry the parasite are most successful when conditions are both wet enough that the mosquitoes can breed and warm enough that the mosquitoes aren't killed by the winter. In some periods of human history and prehistory, malaria spread to some colder and drier regions, but hesitantly. In such regions its effects have always been patchier and less frequent (and, much later, easier to control). In general, over the last ten thousand years, when people have moved to cold or dry places, they have escaped malaria. In some countries, the predictability of this escape has led cooler regions within those countries to become enclaves of elites, elites escaping the parasite by moving outside its preferred niche. Similarly, people living today in countries outside the world's malaria zones continue to take advantage of escape from this parasite. For most of the last ten thousand years, this escape from malaria alone, to say nothing of escape from other parasites, is likely to have been associated with a longer life expectancy and reduced infant mortality. If you are living in a malaria-free zone today, you are very likely to be benefiting from the absence of malaria; you are benefiting from the consequences of the law of escape. Falciparum malaria is just one of hundreds of parasite species with a largely tropical niche. The details of the biology and hence the niche of each of these species is different, but their stories are united in that if one lives in geographic areas outside their niche, one experiences escape.

People experience geographic escape from the effects of parasites by living beyond the niches of those parasites. This can occur in two different ways. I introduced the niche as a singular concept in Chapter 2. However, in reality, each species has two

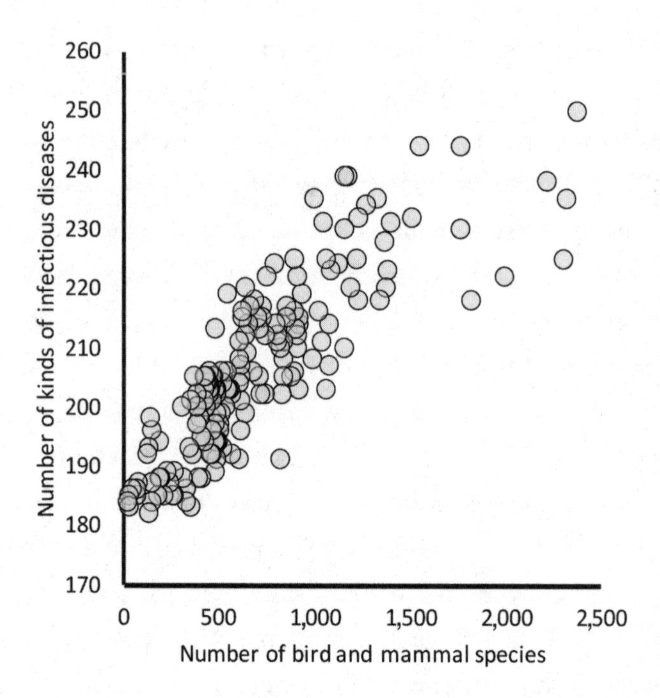

Figure 4.1. The number of kinds of infectious diseases caused by parasites, be they worms, bacteria, viruses, or other taxa, as a function of the diversity of birds and mammals in a political region. Places with more kinds of birds and mammals have more kinds of diseases because the same processes that lead to the evolution of more kinds of birds and mammals also lead to the evolution of more kinds of parasites, the causal agents of infectious diseases.

niches, a fundamental niche and a realized niche. The fundamental niche of a species describes the conditions in which it *can* live and also, often, the geography of those conditions. The realized niche describes the subset of those conditions and geographies where the species is actually found. The realized niche of a species might be smaller than its fundamental niche, if another species prevents it from colonizing a particular realm. But the more mundane context in which fundamental and realized niches differ is when a species simply fails to arrive in a particular place. For example, it is entirely possible that Antarctica contains conditions in which a polar bear might thrive. Antarctica is probably part of the fundamental niche of the polar bear. Yet actually getting to Antarctica from the Arctic is a long swim, so Antarctica is not part of the realized niche of the polar bear.

The distinction between fundamental niches and realized niches is germane to a consideration of escape. Species, including humans, can escape their enemies by moving outside those enemies' fundamental niches. Today, Europeans experience such an escape with falciparum malaria, which can easily arrive in Europe but is more readily controlled when it does because of the biology of the mosquito that carries the parasite and that of the malaria parasite itself. People can also, however, escape by moving to regions that are within the fundamental niche of a particular enemy but not yet part of that enemy's realized niche. They can go to places where their enemies have not yet arrived and, in doing so, escape. Such escape has been important through human prehistory and history, but it is invariably ephemeral. It lasts only so long as the enemy in question fails to colonize all of its fundamental niche. Eventually, our enemies catch up, a reality of great relevance to the future we confront.

ONE OF THE best-studied escapes in human history is the escape that occurred when some human populations began to move from Asia into the Americas over a land bridge that connected the two continents. The land bridge appeared during a period of extraordinary cold, when so much water was frozen in glaciers that sea levels lowered and uncovered a passage. When peoples moved across this land bridge, their movements had a complex effect on the geography of the niches available to the parasites that rely on humans. On the one hand, by colonizing new regions, these peoples made those new regions potentially a part of the niches of many human parasites. However, the fact that these peoples traveled through the cold frozen North first, to get to the rest of the Americas, had a peculiar effect on parasites. Intestinal parasites whose eggs require warmer grounds, such as hookworms, might have been prevented from surviving when the first peoples of the Americas were moving through the far north. Also, parasites that relied on more tropical conditions or vectors would have been left even farther behind. Some parasites of humans that tend to do best in denser human populations, such as the bacteria that cause tuberculosis, shigella, and typhoid, were also left behind.[7] They may have been left behind by chance. The populations that migrated happened to lack these parasites. In this context, there was the potential for the first peoples of the Americas to escape nearly all human parasites not only while living in the far north but then also as they moved farther south. For the most part, this appears to have happened.

I say "for the most part" because it is far harder to fully escape parasites than it might seem. And once a parasite arrives, it can often spread quickly through a human population, as we have recently seen with the virus that causes COVID-19. The virus that causes COVID-19 might have stayed in China, even after having evolved so as to be able to infect humans. The world outside China might have escaped the worst of its consequences. Instead, unfortunately, the virus left China and quickly colonized the world. According to the research of Brazilian parasitologist Adauto Araújo and his close collaborator, University of Nebraska parasitologist Karl Reinhard, something similar appears to have happened in the Americas many thousands of years ago.

Araújo and Reinhard have spent their lives studying the parasites found in mummies and other remains of the peoples of the Americas before the arrival of Europeans. What they have found is that those remains included many species of parasites that could not have survived a journey with people traveling on foot across the far-northern land bridge. The conditions on the land bridge were outside their fundamental niche; they would have died in the cold. On its own, this is a remarkable discovery, and one more piece of evidence that some of the first peoples of the Americas came not over the land bridge but, instead, by boat. (Whether the boat—or those boats—crossed the Pacific or traveled along the coast from north to south is not yet known.) But even more amazing than the evidence that some species of parasites took such a journey is the number of species that did. The list includes hookworms, wireworms, whipworms, and ascarids, to name but a few.[8] A strain of tuberculosis may even have come on this journey (or perhaps these journeys).[9] Having arrived in the Americas, each of these species then colonized all or nearly all the human populations within the geography of the American part of their fundamental niche.

When those parasites spread throughout the Americas, they decreased the extent to which the first peoples of the Americas had escaped from the enemies of humans present in Africa, Europe, and Asia. (This was to be a kind of foreshadowing of the sort that is only clear in retrospect.) Importantly, however, not all species of parasites made the shipborne journey. Many had still been left behind. For the first Americans in the Amazon rain forest, for instance, there was no yellow fever, no schistosomiasis, and no falciparum malaria. But there was also something else.

In Europe, Asia, and Africa, as humans began to tinker more and more with nature during the early days of the great acceleration, they created many new sets of ecological conditions. Species that had long been rare became common. This was true, for example, of many domesticated animals, including pigs, goats, cows, sheep, and chickens. In addition, humans began to live, at least in some regions, in ever-denser settlements. One of the things that disease ecologists wholeheartedly agree upon is that this combination of circumstances is ideal for the evolution of new parasites and the diseases they cause. To return to a concept from Chapter 2, big human populations are like giant islands of habitat from the perspective of parasites. The animals living alongside humans offer an opportunity for many parasite species to colonize those islands. This is just what happened. In large settlements in Europe, Asia, and Africa, new parasites evolved the ability to live on humans and to spread from human to human. In very dense populations, an entirely new kind of parasite even evolved, parasites that spread through the air from person to person. The diseases whose parasites evolved in response to the great acceleration and its associated increases in human population densities and human impacts on ecosystems included influenza, measles, mumps, plague, and smallpox, just to name a few.[10]

Like populations in Europe, Asia, and Africa, populations in the Americas would also experience accelerating rates of population growth. But slightly later. And for reasons that remain not entirely clear, their population growth was associated with the evolution of far fewer new species of parasites.

Ultimately, populations of the Americas experienced an escape not just from one or two parasite species but, instead, tens or even hundreds, some of them very old, some of them new.

Similar escapes would occur, to varying extents, in migrations to islands large and small around the world. People built boats, rigged sails, and paddled and steered themselves away from demons both old and new.

THE ESCAPES OF human bodies from predators and parasites would be repeated with crops. Independently in a half dozen places on Earth, humans figured out how to domesticate crops, cordoning off more of the green world for themselves. Then they also began to move those crops to conditions that were slightly drier than the crops' native conditions. The regions to which domesticates were moved were not the places with the climates or soils best suited to the crops; they were, instead, the places where the crops were most needed by humanity. Perhaps by chance, those also happened to be regions where the crops were sufficiently outside the niches of some pests and parasites to experience some escape. Then humans began to move from region to region on ships.

The movement of humans on ships had two consequences. Human populations moved to new geographic realms and achieved new bodily escapes. Escape to Madagascar. To New Zealand. To nearly every far-flung place. But humans also moved their crops. Crops from South and Central America, for instance, were moved to the Caribbean. Crops from Africa made their way to southern Europe. The crops too escaped; the effects of this escape were greatest when crops were moved to entirely different biogeographic regions.

Over hundreds of millions of years, the relative isolation of different land masses led the animals, plants, and even microbes of different regions to become different. The more separated two regions were, the less likely it was that any species could move between them. Species separated by geography diverged. The more time passed, the more they diverged, so that eventually different regions had very different species. Hummingbirds are found only in the Americas. So too the ancestors of tomatoes, potatoes, and chilies. Tree kangaroos are found only in Australia and Papua New Guinea. So too the ancestors of bananas. Apes are found only in Africa and Asia. Superimposed on these divergences are later movements. In some cases, land masses collided, mixing the species from one land mass with those of another. In other cases, individual species dispersed from one land mass to another: picture two monkeys floating across the ocean on a big log. It is thought that it was via such a journey that primates arrived in the Americas. The differences in the biotas of different land masses that result from this mixture of isolation, tectonics, and dispersal allow ecologists to group land masses into biogeographic regions. For example, most of North America falls into the Nearctic biogeographic region and contains very different species than does the Palearctic biogeographic region, which includes most of Europe and Asia.

As crops were moved from one biogeographic region to another, not only did they often escape their ancient pests and parasites, but they also moved into regions where the relatives of their ancient pests and parasites were also absent. Such moves offered crops a new and more complete escape. The speed with which crops were moved and escaped accelerated when Europeans arrived in the Americas. Chili peppers moved, with the Portuguese, to places like India and Korea, where they would

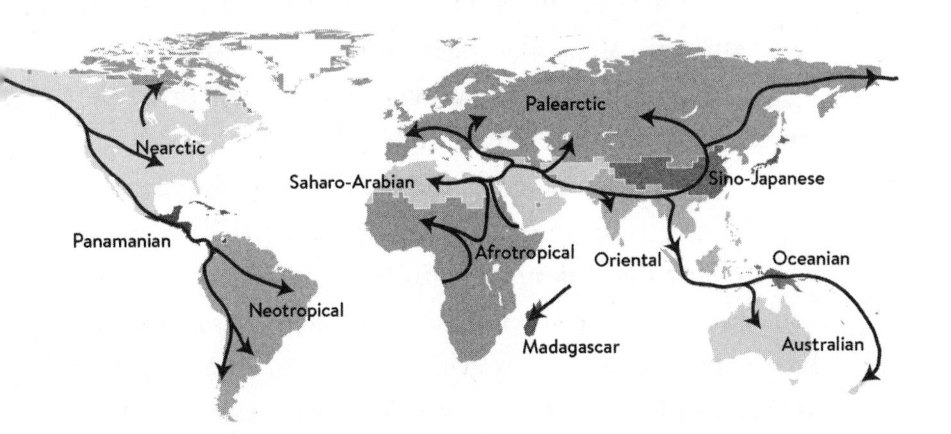

Figure 4.2. The biogeographic regions of Earth, based on the species of amphibians, birds, and mammals present in different regions. Regions that are very different from each other are delimited by white lines and different shades. Lines on the map note one potential trajectory of our species, *Homo sapiens*, as it moved around the world fleeing its parasites and predators. This map was created by Lauren Nichols based on a map in Holt, Ben G., et al., "An Update of Wallace's Zoogeographic Regions of the World," *Science* 339, no. 6115 (2013): 74–78.

become so ensconced in culture and culinary ways that they are now thought of as native. Tomatoes eventually moved to Europe. Potatoes moved from the Andes to Ireland.

With all of this movement, it was inevitable that, in addition to creating opportunities for escape, humans also created opportunities for parasites and pests to spread and thus to colonize the geographic entirety of their fundamental niches. When hosts escape parasites or predators, ecologists call the resulting escape "enemy release." There is no good word for the moment when our enemies find us again, perhaps because no words can readily describe how horrible that moment can be.

When Europeans arrived in the Americas, they brought with them some old parasites that Native Americans had escaped by moving to the Americas. They also brought new parasites that had evolved in the context of large cities in Europe, Africa, and Asia. European ships were filled with every ill to which flesh is heir. The result of the spread of these ills was death on a scale that has seldom been seen. Tens of millions of Native Americans died in what has come to be called the Great Dying. The ancient cities of the Americas collapsed. Populations relocated. The devastation was so great that colonists began to imagine that the Americas had never been very populated. They came to see the ruins of homes and civilizations as evidence of a lost people rather than as the consequences of the comingling of disease and genocide.[11]

Later, parasites from the crops of the Americas caught up with crops that had been moved. The arrival of the potato blight in Ireland reunited the potato with an ancient enemy from which it had escaped. The famine that ensued led to the deaths of a million Irish and the movement of another million to other countries.

TODAY, IN MANY countries the health and well-being of people and the yield of crops still depend on escape of two kinds. The first is the result of species of parasites and pests whose realized niches remain smaller than their fundamental niches. The second is experienced by human populations and crops that live and grow in conditions outside the fundamental niches of their parasites and pests. Both of these kinds of escape are now threatened by global change, the first by the ways we have connected the world through our transportation networks, the second by climate change.

A harbinger of the consequences for escape of the ways we have connected the world could recently be seen when the cassava mealybug caught up with cassava. Cassava is native to the tropical Americas but was introduced to tropical Africa and Asia. In much of tropical Africa and Asia, cassava, released from its enemies, became a primary source of sustenance. To many people of Africa,

Asia, and the tropical lowland regions of the Americas, it is what the potato was to the Irish just before the potato famine.[12]

Then, in the 1970s, cassava came under threat. A new mealybug (kin to the aphid) arrived on cassava plants in the Congo Basin in Africa. The mealybug was accidentally introduced by well-meaning researchers trying to bring new varieties of cassava from the Americas to Africa. It was a devourer. It could kill acres and acres of cassava fields in a year, destroying them from edge to edge. And if it continued to spread at the same rate as it was spreading in the Congo Basin, it would be across Africa within a few years. It would be across Asia in a few more. Nothing seemed able to stop it. The mealybug was growing with abandon. Its populations were growing without any pressure from pests or parasites. It had left all the species that had evolved to prey on it behind in its native range. It had escaped.

One possibility for stopping the mealybug was to go to where the mealybug was native, find whatever insect or parasite kept it in check there, and release that enemy where the mealybug had been introduced. Such biological control would be a long shot. The trick was to figure out what was eating it back home, bring those species to the Congo Basin, raise many individuals of those species, and release them.

To find the enemies of the mealybug, one would have to know where the mealybug came from. No one did. In the absence of knowing where the mealybug was from, one could benefit from knowing where the relatives of the mealybug were from. No one knew which species it was related to, much less where they lived. In the absence of knowing where its relatives were from, one might go to the place where cassava was first domesticated (where its pests and parasites and their pests and parasites might be most common). No one had studied the geographic origin of cassava in much detail. And so, left without any good options, Hans Herren, a scientist too young to know better and young enough to try, started to search. Herren started in California and moved south. One field, one war zone, one hardship to the next. In Colombia he found the mealybug, only to realize it was a different mealybug.[13] One of his friends named it after him, and he kept traveling.

Herren never found the mealybug, but he told his friend Tony Bellotti about his quest. Bellotti happened to be going to Paraguay to sit down with his soon-to-be ex-wife and sign divorce papers. Bellotti had reason to look for distraction. And when he was looking for distraction, he found the cassava mealybug in its native range, in Paraguay.[14] Herren, Bellotti, and others then discovered a wasp that laid its eggs in the bodies of the cassava mealybugs in Paraguay. They took a dozen of those wasps to a quarantine lab in the UK (where an accidental escape would be less likely to prove problematic). Then, after detailed studies of the biology of the wasps, they took their progeny to West Africa where, against the odds, they found a way to turn a few wasps into hundreds of thousands. They released the hundreds of thousands of wasps and, amazingly, the wasps and their progeny spread across Africa, destroying the mealybugs and saving the cassava crop for hundreds of millions of Africans.[15] The same story would later be repeated in Asia.

A small group of scientists, each of them an expert in some obscure facet of the biological world, saved millions of people from hunger. Those scientists are heroes because of their willingness to obsessively search the wilderness of unknown species for a needle—or a wasp, as the case might be—in a haystack. What is perhaps even more amazing, from the perspective of our knowledge, is that after the scientists found the mealybug and realized, in finding the mealybug, that there are probably many species of related mealybugs able to attack cassava (and many species of related wasps that can kill those mealybugs), no one ever went back to study those other mealybugs. Or wasps. Or any of the other species that live with cassava where it is native. Not in any real detail anyway. Those studies will have to wait until the next disaster. We are reminded of the scale of the unknown by near tragedy and actual tragedy. We forget about the unknown in the calm wake of near tragedy and the sorrowful quiet of real tragedy. We forget at our own expense.[16]

The scientists don't forget. They write papers announcing what needs to be done. They give talks announcing what needs to be done. They write a few more papers, discarding their scientist's language for plain spoken alarm. Then, when no one listens, they get back to work doing what they can on their own. So many species attack our crops and so few scientists study those species that we move from disaster to disaster. Sometimes the scientists save things in the nick of time. Sometimes they don't. Meanwhile, there are many hundreds of species of crop parasites that have not yet arrived everywhere they can live.

Today, the sidewalls of car tires and the entirety of airplane tires are made from latex that flows from trees of the species *Hevea brasiliensis*. These trees grow wild in the Amazon rain forest but cannot be grown on plantations there because they are very susceptible to pests and parasites. As a result, almost all the rubber in the world comes from plantations in tropical Asia. There, it grows, having escaped its pests and parasites. But it is only a matter of time before those pests and parasites catch up, and when they do, it has been estimated that the entirety of the global production of rubber could be wiped out within a decade.[17]

Many of our crops thrive today thanks to escape. Many human populations thrive today because of escape. These escapes have occurred in the context of the details of human history but also the details of the geography of pests and parasites. Importantly, that geography is changing.

IN ADDITION TO being threatened by the movement of species around the world via transportation networks, the escape of human populations and of crops is also threatened by the mixture of such movements and climate change. Here, let us consider the case of the *Aedes aegypti* mosquito.

The viruses that cause yellow fever and dengue fever are both carried from the blood of one person to that of another in the delicate bodies of *Aedes aegypti* mosquitoes. When the first peoples arrived in the Americas, both the two viruses and the mosquito were absent. For more than ten thousand years, they remained absent. The people of the Americas lived without fear of yellow fever or dengue fever. Then, eventually, the *Aedes aegypti* mosquito arrived. It appears to have been introduced to the Americas on slave ships and then spread via the corridors engendered by networks of roads, rivers, and railroads. The yellow fever virus appears to have come, like the mosquito, in slave ships, riding in the bodies of enslaved people. Later, the dengue virus found its way to the Americas too, traveling from Asia. The yellow fever virus, the dengue virus, and the mosquitoes they ride in now live throughout the sun-warmed latitudes and sun-warmed cities of the Americas. As climates change and as cities become larger and more connected, they will, to varying extents, continue their spread.

Aedes aegypti is often called a "domestic" or even "domesticated" mosquito species because it is most likely to thrive around people. Cities have the habitat the mosquito requires: small patches of water in old tires, gutters, and the like. In addition, cities are often warmer than their surrounding habitats, and *Aedes aegypti* is a tropical mosquito species; it thrives on warmth and is killed by cold winters. However, in some regions that tend to be too cold for *Aedes aegypti*, it nonetheless persists thanks to warmer urban conditions. One population of *Aedes aegypti* appears to have become established, for example, in Washington, DC. It lives near the National Mall, and when the mall is cold in the winter, it hides out underground in any of the many structures built by humans beneath the nation's capital. Most species will struggle to keep up with climate change, but the warm places in cities will allow heat-loving, urban-dwelling species to move north in advance of the change.

That *Aedes aegypti* is spreading along the corridors of cities, leaping northward out of the tropics like a flame and surviving through the winter, is a deadly problem that will affect people throughout large parts of the United States and also in other parts of the world. The conditions required for the yellow fever virus and the dengue virus to persist are subtly different from those required by the mosquito. But once the mosquito has become established, it is far easier for either virus to gain a foothold as well. Studies in which scientists use everything they know about the biology of the *Aedes aegypti* mosquito to predict its distribution in the future suggest that much of the eastern United States will be coping with the mosquito and the risk of dengue fever epidemics in the next few decades. Whether they will also be coping with yellow fever will depend on complexities of the interactions between the dengue virus and the yellow fever virus (via the human immune system), the distribution and abundance of *Aedes aegypti*, the distribution and abundance of another mosquito introduced to the United States, *Aedes albopictus*, with which *Aedes aegypti* competes, and the distribution of the other mammal species in which the yellow fever virus lives. What we do know is that much of the southern United States is going to be coping with new problems associated with *Aedes* mosquitoes. Those problems will include some complex mix of the dengue virus and the yellow fever virus, but also the viruses that cause chikungunya, Zika fever, and Mayaro. The bigger truth, though, is that by connecting the world and altering the climate, we are shifting the parasites that can live in each and every region *and* affecting how those parasites can move within the geography of their fundamental niches.

SUPERFICIALLY, THE CHALLENGES of predicting the fate of parasites in the future are very similar to those of predicting the fate of birds, mammals, or trees. However, the added challenge with parasites is that they tend to have more complex life cycles than do birds, mammals, or trees. In addition, parasites also tend to be more poorly known than are vertebrates or plants (thanks in part to the law of anthropocentrism). As a result, if you consider each parasite species, one by one, it is easy to get overwhelmed both by the details and by how much is unknown. The data depicting the distribution of different parasite species are, with a tiny handful of exceptions, terrible. As a colleague and I recently showed, we know far more about the geography of bird species, even very rare warblers, than we do about the geography of even relatively common parasites of humans.[18] This is true even though there are many fewer species of parasites that affect humans than there are bird species. What scientists have tended to do, when confronted with this reality, is to focus on a few of the worst parasites; we know a lot about where malaria will spread, for instance, and nearly as much about the potential spread of dengue. But this leaves out most species, and when we remember that the parasites that will be moving include not only parasites of humans but also those of crops and domestic animals, the task becomes more daunting and the hope of seeing clearly even more remote. Fortunately, there is a rule of thumb approach that can be helpful.

Climate scientists have gotten ever better at predicting the future climate of different regions, given different scenarios with regard to human behavior. As a result, we can take a particular region of interest—for example, New York or Miami. We can then examine the future climate of that region, and we can map out what other regions currently have similar climates. The parasite species found in those climate-similar regions then offer a reasonable estimate of at least a subset of the species that might live in, say, New York or Miami in the future. Think of this as the parasite sister cities approach.

The parasite sister cities approach allows me to estimate the parasites most likely to be able to survive in any city in the future under different climate scenarios. Climate scientists think of the future, just as modelers of urbanization do, with regard to scenarios, each of which reflects a set of human behaviors and how climate will respond to those behaviors. Climate scientists have no special expertise in predicting human behavior, but they have honed their ability to understand how climate will respond to different sets of behaviors, different scenarios. Each scenario portrays a set of human behaviors and decisions, the emissions of greenhouse gases that result from those behaviors and decisions, and the changes in climate that result from those greenhouse gases. These scenarios do not, in and of themselves, specify what we should do; instead, they describe what will result given different things we might collectively do.

The scenarios differ with regard to how much humans reduce greenhouse emissions and, consequently, minimize the extent of climate change. The scenarios differ in their degree of optimism about our global collective human behavior, with the more optimistic scenarios assuming that we rapidly change our ways and reduce our greenhouse gas emissions. Some of the scenarios are no longer possible; we have already failed to achieve the change necessary to make them possible. The most optimistic of the scenarios that are still possible is called RCP2.6. To be on track for this scenario, we would have begun reducing our global emissions of the greenhouse gases that contribute to climate change by the year 2020, the year before this book was published. We would have reduced our greenhouse gas emissions by 7.6 percent that year and then, looking forward, we would need to continue to do so in each subsequent year until 2100, at which point the greenhouse gas emissions by humans would need to reach zero and stay at zero. Zero. This RCP2.6 scenario is very unlikely.

The second scenario, RCP4.5, is not as hopeful, and yet it still requires radical change that needs to begin immediately. In the second scenario, we stop *increasing* our emissions of greenhouse gases by 2050, despite projected increases in the number of people on Earth. In other words, our individual emissions must actually decline rather dramatically in order to hold our collective emissions constant. Achieving this scenario requires a rapid shift to renewable energy, a shift away from the consumption of meat, and global decreases in the number of children people have, among other changes. If you are living a lifestyle even remotely similar to the life you were living ten years ago, with regard to diet, travel, daily transportation, or heating and cooling, you are unlikely to personally be on the trajectory required for this path. RCP4.5 requires radical change and yet nonetheless yields about two additional degrees Celsius of warming, globally.

The third scenario: we continue to use fossil fuels as we have been using them. The name of this scenario is RCP8.5, or the business-as-usual scenario. RCP8.5 yields a whopping four degrees of climate change by 2100. In my experience, people who study climate change are planning for this last scenario in their own daily lives. At work, they write about the RCP2.6 path and how to get on it. At home, in their free time, they fight to get their communities on the RCP2.6 path. But when their day is over and they are sitting on the couch, they make choices that reflect their worry that we are probably on the RCP8.5 path. They look online at properties in, say, Canada or Sweden. They talk to real estate agents and ask questions like, "Does it have year-round flowing water?" They have conversations with their partners about which countries have stable governance and won't have malaria. With insider information and disposable income, they are preparing in advance to flee. Here, I am reminded again of the story of the ark. Noah, upon being told that great floods would cover the terrestrial world, tried to tell people he knew. No one listened.

These three scenarios are among a larger number of related scenarios that were laid out by the Intergovernmental Panel on Climate Change (IPCC) in 2014. The IPCC decided to offer a portfolio of scenarios rather than a specific prediction because laying out such a portfolio makes our choices clearer, but also because it is far easier to predict how a certain level of emissions will affect climate than it is to predict collective human choices and behaviors. (Since that time, a new set of scenarios has been developed that make slightly different assumptions about human behavior. They have new names and new details but yield remarkably similar predictions to the scenarios I've listed above.) Climate scientists cannot say whether we will choose to stay on the business-as-usual trajectory (RCP8.5) or, instead, radically reimagine the ways we live (RCP4.5). Our choice is with regard to how much we change and how much climate changes us.

It was in light of these scenarios that, a few years ago, my colleague Matt Fitzpatrick developed a tool that allows people to see which city in North America their city will be most like in the future (by 2080, give or take) given the RCP4.5 and RCP8.5 scenarios. Matt doesn't call it the parasite sister cities approach, but he doesn't mind if I do—at least, I hope he doesn't.

Figure 4.3 depicts the future of a handful of cities in light of Matt's results. Matt focused on the RCP4.5 and RCP8.5 scenarios. Each line from a particular city connects to the places with current climates most like the future climate of that city in 2080. The left panel shows the results for the RCP4.5 scenario, the right panel for the RCP8.5 scenario.

The lines on the maps offer a very direct measure of the parasites of our future. We can focus on Miami, Florida, for some context. Miami, in scenario RCP4.5, becomes similar in climate to the subtropical parts of Mexico, which is to say very hot and seasonally wet. In scenario RCP8.5, it becomes more like tropical Mexico. Or at least those parts of Miami that are not underwater become more like tropical Mexico.

What the match between future Miami and parts of Mexico tells us is that in the future Miami will be within the fundamental niche of most species that live in subtropical Mexico (RCP4.5) or tropical Mexico (RCP8.5). This affects the potential wildlife of Miami—picture a future of monkeys and jaguars. Species from the habitats in Mexico in which monkeys and jaguars now live will need to make their way to Miami. They will need to track their niches all the way there even though vast stretches between

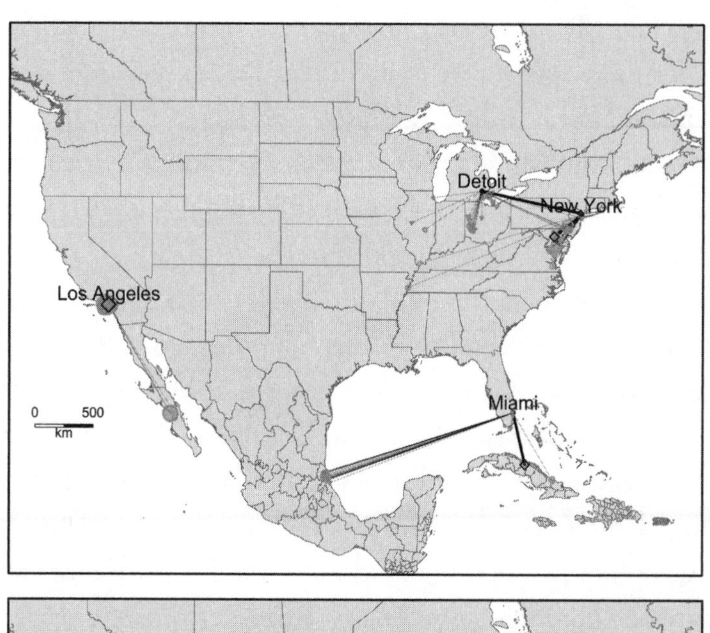

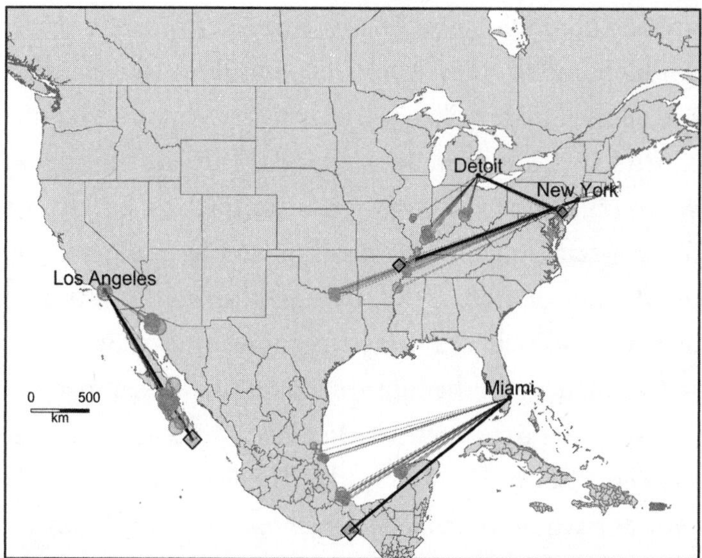

Figure 4.3. For several cities, the "sister cities" that most closely match their future climate under RCP4.5 (*top*) and RCP8.5 (*bottom*) scenarios. Different lines indicate the results of different climate models. The smaller a shape at the end of a line, the more similar a match. The larger a shape, the less similar the match. The online version of this map allows you to choose any city in the United States and chart its future. Diamonds and dark lines depict the average results of all the models.

Mexico and Miami are very hot and dry. Their route is a difficult one. For these species, we don't need walls between Mexico and the United States. Instead, we need corridors of forest stretching all the way from Mexico to Florida, and beyond. We need an ark of habitat; much of it won't be perfect for the species that need it, but we'll have to hope it is good enough.

Meanwhile, parasites find themselves with no shortage of boats, planes, highways, and other transport. And the parasites of Mexico are many. The climates of tropical Mexico are suitable for malaria parasites and the mosquitoes that carry them, for dengue fever and yellow fever viruses and the mosquitoes that carry them, for the parasite that causes Chagas disease and the bugs that carry that parasite. In addition, tropical Mexico hosts many parasites of domestic animals and crops that do not yet thrive in Florida. Some of these species of parasites or the species they require, be they mammalian hosts or insect vectors, have already been introduced, accidentally, to Florida. They are waiting, in the very warmest bits of the state, for conditions to be just a little warmer. The same sorts of comparisons can be made with regard to the parasites of crops. And, by going to Matt's website, they can be made for each and every city in the United States.[19] Matt focused on matches to the future climates of different parts of the United States to places elsewhere in North America. But there are also matches elsewhere in the world. There are parasites in Africa and Asia for whom Miami becomes part of their fundamental niches in the future. For such parasites, the challenge of arriving is greater and yet, if history is a lesson, far from insurmountable.

THE PROBLEM WITH escape is that people don't value it when they have it. The majority of people on Earth live in places where neither they nor their crops have escaped from the majority of tropical parasites. But if you live in Miami, you have escaped many of the worst of our enemies. Yet the consequences of those parasites can seem far away and abstract. It is difficult to convince people not to build their houses on sites that will be underwater as the climate changes. It is even more difficult to convince them that they need to plan for parasites that aren't here yet and might or might not arrive within their lifetimes. It is nearly impossible to convince them that they need to plan ahead of the great migrations of parasites, given that the planning that needs to take place is both boring and heavily detailed. Yet we can plan ahead. There are several simple steps.

The first step we can take is to stall. Anything we can do to keep the waves of parasites from arriving benefits many people. It is far easier to keep a parasite from arriving, albeit nonetheless difficult, than it is to control a parasite once it has arrived. We need to monitor for the insect vectors of the worst parasites. We need to enlist the public in monitoring for those vectors. And we need to develop robust public health surveillance systems for monitoring the parasites themselves. In some places, these systems are partially in place. In no place are they yet sufficient. In most parts of the United States, for example, the lag between when a new mosquito species arrives and when it is detected is often a decade. By then, we notice it because it has become common. By then it is too late.

Meanwhile, we also need to prepare our public health systems to be ready to deal with new kinds of parasites. When Mike Gavin, Nyeema Harris, Jonathan Davies, and I modeled the diversity of diseases caused by parasite species around the world, we came to two key conclusions. The first was, as I've already mentioned, that parasites are most diverse where conditions are hot and wet. Climate predicts the diversity of diseases very well. One might have hoped that, given the amount of money humans have spent on disease control, we might have altered the ancient connection between climate and disease. We haven't. This is humbling. Where it is hot and wet, there will be more kinds of diseases caused by parasites. But there was also a second conclusion. The prevalence—which is to say, the proportion of people infected—of the worst human diseases was not explained solely by climate. It was, instead, better explained by a model that incorporated both climate and expenditures on public health. Public health expenditure, in other words, does not often extinguish parasite species, but it can keep them rarer than they might otherwise be. It seems likely that something similar is true for agricultural parasites and pests. The countries and states that will become more tropical in the future should start investing in the infrastructure to control the arriving hordes.

The other option, of course, is to try to flee again. We could, as some have argued, colonize the moon or Mars. As an ecologist, it seems unlikely to me that we could engineer entirely new ecosystems on other planets that we could manage sustainably when we have struggled to avoid destroying the already functioning ecosystems around us on Earth. But for the sake of argument, let's imagine that we could colonize the moon or Mars. Imagine that Elon Musk has a summer house there, with a nice (but sealed in) porch. Imagine greenhouses full of the most delicious plants. Imagine the things we love about our own planet, replicated in some simpler form. Imagine a settlement free of parasites of any kind. In such a scenario, we could escape again. Or rather, a handful of wealthy individuals could escape again. But if the past tells us anything, it is that even this escape would be temporary. For example, researchers recently found plant parasites growing in gardens maintained by astronauts on the International Space Station. Plant parasites are already in space.[20]

CHAPTER 5

The Human Niche

MOST SPECIES ON EARTH WILL NEED TO MOVE WITH CLIMATE CHANGE in order to home in on the conditions in which they thrive. The species that will need to move to survive include rare birds, snails, and parasites alike. This much I've already said. What I did not say is that these species also include humans. In some ways, the diversity of climates and conditions that humans have come to inhabit and turn fleshy as they have fled and explored is astonishing. Our niche seems broad. Since before the invention of agriculture, humans have managed to settle in tundras, swamps, deserts, and rain forests. Innovations have enabled our modern human species to occupy far more biomes and conditions than did any ancient human species. If we zoom in on individual people and their societies, it is these innovations that come into focus. They include the invention of fire and clothing to keep warm, irrigation systems to move water, and the ability to heat and cool buildings. They also include ways of life uniquely adapted to certain conditions.

Pastoralists around the world live in extreme environments by moving seasonally with their animals. Far-northern peoples survive by relying on an extraordinary knowledge of the animals and plants around them, coupled with seasonal migrations, food storage, and novel building techniques. Modern science has figured out how, at least temporarily, to colonize space. Above us now, astronauts may well be eating breakfast, sleeping, or reading.

But if we zoom out and look at humans as a whole and consider not where humans can live but, instead, where *large*, dense human populations have been able to persist, the picture changes.

When we zoom out, the importance of our innovations all but disappears. Instead, the physiological limits of human bodies become more apparent. For example, Chi Xu at Nanjing University in China and his collaborators at Aarhus University, the University of Exeter, and Wageningen University have recently measured the ancient and modern human niche, based on data on the population density of humans everywhere on Earth. If we want to measure what conditions favor human survival, considering population density is a reasonable place to start.[1]

Xu and his colleagues graphed the relative proportions of land on Earth in different climates. Doing so revealed the wide variety of combinations of temperature and precipitation found at least somewhere on Earth, from the very cold and dry to the very hot and wet. It also, however, revealed that some climates are far more common than others, and perhaps also more common than we tend to imagine. Much of the terrestrial Earth is either as cold and dry as the most remote tundra or hot and dry like the Sahara. Xu and his colleagues then examined the subset of these conditions that have allowed for the persistence of dense human populations. They used the same approach that ecologists, including some of Xu's collaborators on the project, use to consider the niches of nonhuman animal species. They studied humans the way they would any other animal, be it a honey bee, a beaver, or a bat.

Xu and his collaborators first studied the human niche during the relatively distant past, six thousand years ago, based on various kinds of archaeological data that have recently been compiled in an online database. Six thousand years ago, a much larger proportion of the global population was composed of hunter-gatherer populations than is the case today. In considering these ancient people, Xu and his colleagues found that they relatively densely occupied a wide range of climatic conditions, but not all such conditions. This can be seen in the top middle panel of Figure 5.1, where the brightest white shows the climatic conditions in which human populations achieved their highest densities six thousand years ago. What you will quickly notice is that ancient people tended to live in only very low densities in very cold regions and hot, wet conditions, while some of the hottest, driest places on Earth were relatively densely inhabited. The highest-density conditions, however, were those associated with moderate temperatures and relatively dry conditions. The "ideal" average annual temperature for ancient human populations, at least from the perspective of density, appears to have been about 13°C (55.4°F), roughly the mean annual temperature of San Francisco in the United States or Florence in Italy. And the ideal precipitation was about 1,000 mm of precipitation per year, which is wetter than San Francisco but similar to Florence. In ancient times, long before air conditioners or central heating, such pleasant climates allowed large human populations to thrive.

In turning from that ancient past to our modern moment, the question is how much we humans might have expanded our niche through technology with our remarkable powers of innovation. The surprising answer is, for the most part, not at all. In the

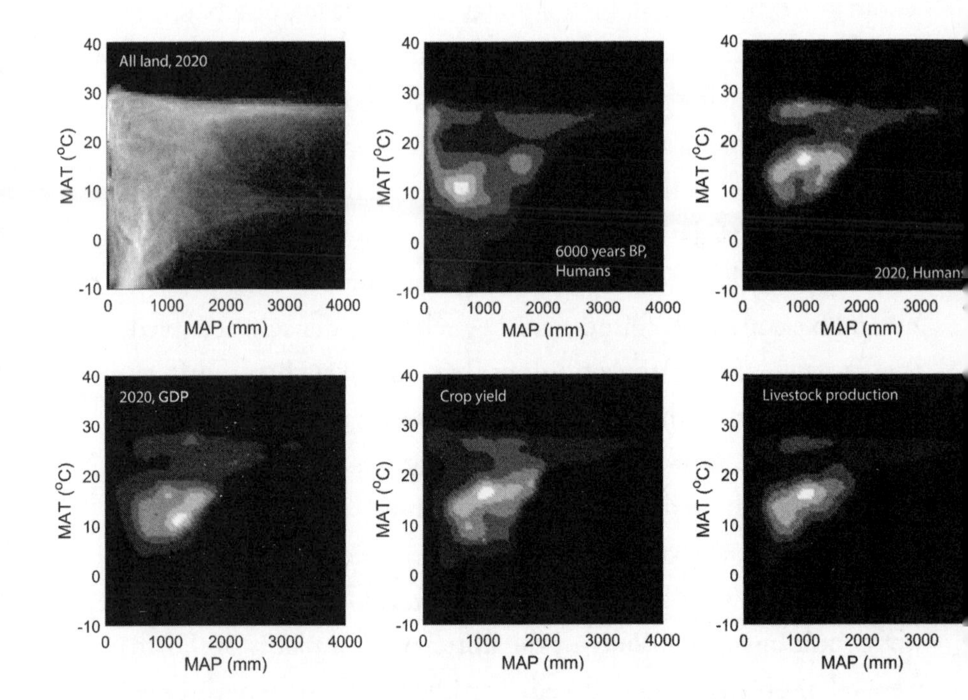

Figure 5.1. The area of land in different climates (*top left*), the number of people in different climates six thousand years before the present (*top middle*), the number of people in different climates today (*top right*), gross domestic product (GDP) as a function of climate (*bottom left*), and crop yield (*bottom middle*) and livestock production (*bottom right*) as functions of climate. In the top left panel, the brighter whites correspond to climatic conditions that cover a larger proportion of the terrestrial earth. In the top middle and top right panels, shades correspond to the population density of humans. The brightest white indicates areas in which population densities are very high, 90 percent of the maximum. The next lighter color corresponds to areas where populations reach 80 percent of the maximum. And so on. For the bottom panels, the brightest white indicates climates in which GDP, crop yield, and livestock production are 90 percent of the maximum. MAP = mean annual precipitation; MAT = mean annual temperature. Figure produced for this book by Chi Xu and Lauren Nichols.

ensuing years, our occupancy of Earth has not spread out more evenly across climates; instead, it has become more concentrated. Despite all our innovations, despite steam power, coal power, nuclear power, air-conditioning and central heating, desalination plants, and all the other shiny bangles of modernity, the human niche has, if anything, contracted.

Six thousand years ago, the peoples who lived in very cold, dry conditions tended to be hunter-gatherers reliant on the fish, birds, and mammals of the far north. Cultural innovations allowed these hunter-gatherers to thrive despite the seasonality of their foods (they fermented foods for preservation), the extreme cold (they insulated themselves and also learned to tolerate what others might not), and the vast distances (in some places, they came to rely on sled dogs). Similarly, six thousand years ago, pastoralists found ways to live in hot, dry realms. They relied on the animals that they pastured (they ate them, drank their milk, and used their skins and meat), on seasonal movements, and on clothing and houses that made the heat bearable. They also simply became accustomed to bearing what others might not.

Today, many of the extreme places these people once occupied are now relatively uninhabited or are occupied at such low densities that they no longer represent much of the global population. For example, fewer people live in parts of the hottest parts of the Sahara today than did six thousand years ago,[2] and those people now represent a much smaller proportion of the global population. Similarly, parts of the tundra are less densely inhabited today than they were six thousand years ago. In their research, Xu and his colleagues conclude that our modern innovations did not expand the human niche, as experienced by the masses, beyond what was possible with the innovations traditional peoples living six thousand years ago had already made. This poses a problem because in the coming years, Earth's climates will become more extreme, hotter nearly everywhere, much drier in some regions and much wetter in others. Given the more extreme future that awaits us, it seems important to understand why extreme climates might pose challenges for human populations in the first place.

Why might more extreme climates negatively affect humans even when we spend much of our time indoors in temperature-controlled buildings? This is a very important question, yet it has not received much attention from ecologists or even anthropologists. Intriguingly, it has been best studied by economists. A number of years ago, a small group of climate change economists, including Solomon Hsiang and his collaborators and mentors, set out to explore how two aspects of human society are affected by climate. The first, unsurprisingly, given their field, was the gross domestic product (GDP) of countries. The second was violence. I'll consider the violence first because the links between climate and violence are more direct than those between climate and GDP.

When Hsiang was a graduate student, the effects of climate on economics weren't considered particularly pressing in his field of study. In part, this was for reasons of history. In the 1950s and 1960s, the field of anthropology revolted against an idea called environmental determinism. Soon other humanist fields, including economics, followed suit. Environmental determinism is the notion that human societies, like ant societies, are influenced by the environment. Humanists were reacting, in part, with good reason, to versions of determinism that reinforced racist and colonial ideologies. Yet, Hsiang felt that humans, nonetheless, still do respond to the biological and physical world. He was young enough that he didn't, in his own telling, know about all of this history. He was just interested in climate and economics and humans, so he started to study them as a graduate student at Columbia University.

Hsiang's PhD led to a series of papers on the effects of cyclones on economies. Following this work, he moved to Princeton University as a postdoctoral researcher and began broader research on climate change and societies, research that culminated in a single comprehensive opus that he published in the journal *Science* while he was a postdoctoral researcher at Princeton.[3] The paper, coauthored with economists Marshall Burke and Edward Miguel, both at the time at the University of California, Berkeley, was, as they would put it, "the first comprehensive synthesis" of what was known about climate and human societies. It was to be a statistical synthesis. Statistics would be the magnifying lens through which the team stared at humanity. Previous studies had considered links between changes in temperature and individual societies but not with a holistic set of analyses. Hsiang, Burke, and Miguel sought to combine these efforts to see the big picture.

The approach of Hsiang and colleagues was complementary to and independent of the approach of Chi Xu and his colleagues. Whereas Xu focused on how population densities of humans relate to climate across space in particular slices of time, Hsiang focused on the relationship between human societies and climate at particular points in space across different times.

What Hsiang, Burke, and Miguel found was that human societies, when faced with rapid changes in climate, especially relative to the conditions in which large human populations are most likely, nearly always suffer. That suffering seems especially pronounced where climate changes in ways that lead it to be more extreme than the conditions circumscribed by the human niche, and that suffering has a common element, visible across time and circumstance: violence.

Changes in climate in general, relative to the human niche, and increases (and, more rarely, decreases) in temperature in particular, both tend to increase violence. People are more likely to inflict violence on themselves when climate changes. Suicide and attempts at suicide increase with warming temperatures. People are more likely to inflict violence on other people. In the United States, domestic violence and rape both increase with increasing temperatures. Violence of individuals against groups also increases with increases in temperatures. This includes retaliation by baseball pitchers against members of the other team (which increases with temperature) and violence by individual police officers against the public (which increases with temperature).[4] It also includes violence of groups of people against other groups of people. Studies that Hsiang and his colleagues considered found that intergroup riots in India increased with increasing temperatures, as did political and intergroup violence in East Africa, as did intergroup violence in Brazil. The list continues. Importantly, on top of all of this, the violence associated with war and the collapse of societies also increased with increasing temperatures, whether in the ancient Mayan Empire, the ancient Angkor Empire, the dynasties of China, or modern cities, states, and countries.

The violence that Hsiang, Burke, and Miguel saw with changes in temperature, as well as with changes in precipitation, resulted from shifts in conditions relative to the human niche. It appears that the more dissimilar conditions are from the ideal human niche, the more people suffer and the more violent people become. Imagine a map of the world on which the places at the edge of the human niche, as measured by Chi Xu, are displayed. Now, superimpose climate change across that map. The research of Hsiang, Burke, and Miguel suggests that violence is likely to be most common in the geographic areas that are currently marginal, with regard to their climate, and worsening. Prompted by this realization, I reached out to Xu to make such a map, and he did. It was clearly visible on his map that the hot spots of global violence, at least the sort of violence that occurs between groups of people, appear disproportionately in two sets of climatic conditions: one in climates that are extremely hot (and typically getting hotter), and the second in regions that are hot and relatively dry, regions with enough rainfall for agriculture in a good year but not a bad year. Regions with the former conditions include parts of Pakistan. Regions with the latter conditions include northern Myanmar, the border between India and Pakistan, and parts of Mozambique, Somalia, Ethiopia, Sudan, Niger, Nigeria, Mali, and Burkina Faso, all of which are experiencing waves of violence.

As conditions shift relative to the ideal human niche, and especially as temperatures increase, a number of things happen that can precipitate the sort of violence that Hsiang, Burke, and Miguel found in their study and that is being observed today around the world. It has been hypothesized that as temperatures increase, bodily consequences might be felt disproportionately in human brains, where they are associated with impaired decision-making and, specifically, with impaired ability to control impulses. Increasing temperatures could affect decision-making even if average temperatures aren't so very high as long as daily maximum temperatures are high. The bodily stress of heat, some have suggested, might make the mind begin to work less rationally than it might otherwise. The ancient part of the brain takes over, the brain of fear and rage and impulse, the lizard brain, all chemicals and consequences. Even in relatively cool regions, this has the potential to happen on hot days. In hot regions, well, it has the potential to happen on many days.

In one experiment, psychologists drove a car up to a stoplight and then, when the light turned green, they waited. They sought to see, under different circumstances, how long it took the person behind them to get mad enough to honk. The hotter it was, the more honking they found. The relationship was linear, and it was even more pronounced if the driver had the car window open and hence was experiencing the full force of the outdoor temperature. At higher temperatures people were more likely to honk and honked for longer. As the authors of the study put it, "at temperatures over 100 degrees F, 34% of the subjects who used their horn leaned on it for over 50% of the total green light interval. In comparison, no subjects below 90 degrees did so." Miraculously, even though this experiment was conducted in the United States, none of the psychologists was shot.[5]

In another study, a group of participants was left in a room, which was then warmed to uncomfortable temperatures. As the temperature went up, the participants began to argue more than they had at cooler temperatures. Each time this experiment was repeated, the results were similar. More warming, more arguing and aggression. In one case, a participant even tried to stab another participant with a knife. Other studies have also found that, at least under certain circumstances, as temperature increases, cognitive control, the ability to consciously make decisions, declines.[6]

Similar patterns emerge when we look at what one could call violence against possessions, the spiteful destruction of property. In one experiment, led by Ingvild Almås at Stockholm University along with a large team including Solomon Hsiang and Edward Miguel, researchers had participants in Berkeley, California, in the United States and Nairobi in Kenya complete preference tests and also participate in an online role-playing game used to study human behavior. In the role-playing games, individuals had opportunities to be fair (or fail to be fair), to cooperate (or fail to cooperate), and to trust (or fail to trust). In addition, in one version of the game, called Joy of Destruction, players could choose to destroy another player's winnings. Doing so offered no benefits to the player who carried out the destruction, but it disadvantaged the player who lost the winnings. Such actions were the very definition of spiteful. Almås and her collaborators staged 144 sessions with twelve participants at each session. Half the participants in each session were asked to play the game at a relatively pleasant temperature of 22°C (71.6°F). For the other half, however, Almås and her collaborators increased the temperature of the game room to 30°C (86°F), an unpleasant though not dangerous microclimate. They wanted to know whether tendencies toward fairness, cooperation, and trust would decline at higher temperatures and whether spiteful behavior would become more common.

Almås and her collaborators found that most of the economic decisions people made while playing the game at the higher temperature were similar to the decisions they made at the lower temperature. Temperature did not, on its own, have an effect on the tendencies of individuals to be fair, trusting, or cooperative. Nor did it affect simple measures of cognition. However, in Nairobi (but not Berkeley), the eagerness with which people spitefully destroyed the possessions of others was about 50 percent higher at the higher temperature. Temperature, in other words, sometimes seemed to increase violence, at least spiteful virtual violence, against possessions.

There was also something more. When Almås and her colleagues carried out their experiment in Nairobi, it so happened that they did so at a time when one ethnic group, the Luo, had recently been marginalized by an election that favored the majority ethnic group, the Kikuyu. This marginalization influenced the results of the video-game test. Individuals who were part of the marginalized group were even more likely to destroy other people's things during the video games. If the Luo individuals were excluded from consideration, the effect of temperature on the tendency to destroy virtual property disappeared. In short, temperature increased violence against property via some mix of effects on psychological state and discomfort, but it only did so in the context of power differences and ongoing hostilities between two groups of people.[7]

Beyond the psychological effects of rising temperatures, another explanation for increases in violence with increases in temperature is more idiosyncratic. It relates to the ways in which temperatures influence logistics. For as much as the world can sometimes seem futuristic, many of the most brutal tasks are still carried out by human bodies. Human bodies pick fruit, load trucks, and kill pigs and chickens, and so it is still human bodies on which the global economy depends. Fifty percent of global agricultural production alone depends on the work of small landholders, who do much of their work outdoors by hand. Collectively, those human bodies, with their innumerable arms and legs, are directly susceptible to the effects of temperature. Economists measure this effect by studying the amount of labor supplied per minute by people as a function of temperature. As the temperature exceeds the temperatures at which humans work with most ease, the average amount of labor people supply per minute declines. And when labor supply declines, the effects cascade through society. The functioning of the world economy and local society alike depend on human bodies and minds; they depend on whether or not individuals decide to mop the sweat off their brows and keep working rather than taking up arms. At some point, as Hsiang and his colleagues put it in their paper, "the value of engaging in conflict" rises "relative to the value of participating in normal economic activities."

At pleasant temperatures, billions of often unseen limbs carry us through our days. But as temperatures warm, those limbs slow until, at some maximum, they will move no more. The effects of temperature on the bodily work of humans tends to be greater in poorer countries because in those countries less work takes place indoors and even that which is indoors is less likely to be done with the benefit of air-conditioning. It is easy to imagine how rising temperatures make such work harder and, above some critical temperature, stop it entirely.

Another way temperature has the potential to affect society is through something often termed "policing," which encompasses more than just the work we would recognize as the duty of actual uniformed police. It relates to the ability of the humans who enforce a society's rules to do work outside. When it gets too hot, police don't give traffic tickets. As a result, those who consider this effect contend, people speed more. When it gets too hot, food safety inspectors go out less often. In the meantime, while policing declines amid rising temperatures, societal problems tend to multiply as government funds dry up due to declining tax bases. When policing declines, all those things that were being held in check bubble up.

A final way rising temperatures or other changes in climate push up against the boundaries of the human niche is via effects not directly on humans but, instead, on the species on which we depend. Humans, as I will consider in Chapter 8, are dependent on many thousands of species, but they depend disproportionately on a relatively small number of crops and domestic animals. Chi Xu and his colleagues were able to show in their work that the bounds of the human niche partly reflect the places where our crops and domestic animals thrive and then, when conditions are too cold, too hot, or too hot and wet, fail to thrive.

Returning to Figure 5.1, one sees that the contemporary human niche and the contemporary niches for crops and domestic animals are very good matches, particularly at high temperatures. Above an average annual temperature of about 20°C (68°F), the yields of most of humanity's staple crops decline; so too do the densities of humans. That is, what Chi Xu and his colleagues have really mapped when considering modern patterns in population is not the human niche but, instead, the niche of agricultural peoples. Because the high-density populations in our modern world are only possible in the context of agriculture, the high-density-living niche and the agriculturalist niche are, today, essentially synonymous. They were not synonymous six thousand years ago, in part because even though hunter-gatherer and pastoralist populations lived at low densities, they were a larger proportion of the global total.

Studies have shown that the effects of changes in climate on crops and domestic animals tend to be greatest where high temperatures and low rainfall co-occur (though effects can also occur due to excessive rainfall alone). Once crops fail, food shortages will occur and so too will a variety of forms of instability and violence. In some cases, the instability and violence are concentrated within the parts of countries most affected by climatic change. In other contexts, violence associated with climate-induced crop failures in the margins of the human niche affect countries or broader regions. In 2010, a heat wave in Russia led to a rise in global food prices due to its effect on Russian agriculture. Increases in food prices can lead to massive migration. When migration occurs to cities that are predominantly based on agricultural economies, the effect can be twofold. Hungry rural people and hungry urban people meet in the city. This cascade of events is relatively indirect and yet important. Temperature increases affect crops, which affect the livelihoods of farmers, which triggers migrations to cities, which can lead to social instability. Social instability fells governments.

IF CHI XU and his colleagues are right about the climatic conditions that circumscribe the modern niche, especially the modern niche of agriculturally dependent humans, and if Solomon Hsiang, Marshall Burke, and Edward Miguel are right about the effects of straying outside that niche, then one might predict that the effects of changes of temperature, year to year, should be visible in the kinds of data that economists love to measure each year for economies around the world. For example, one should be able to see the effects of rising temperatures on countries' GDP (the GDP is a measure of the value of goods and services produced in a year). If Xu and Hsiang are right, the GDP of countries should be expected to increase as temperatures (or other conditions) get closer to the optimum for the human niche. Then, as temperatures increase (or decrease) from this optimum, GDP should decrease for all the same reasons that violence increases; such declines in GDP might even be an early warning signal, a portent of worse dangers yet to come.

Until recently, no one had checked. So Hsiang, Burke, and Miguel got their team back together to gather the necessary data. They then considered how changes in temperature, year to year, in individual countries affected the GDP of those countries. Their results were entirely in line with Xu's results. Just as Xu had, Hsiang, Burke, and Miguel identified about 13°C as the optimum average annual temperature for economic output. They found that when temperatures were below the optimum for the human niche, increases in temperature consistently increased GDP. Think about Denmark, Scotland, or Canada, where a warmer-than-average year might increase the time during which work can be done outdoors and also, at the same time, agricultural yield.

Conversely, in countries where the average annual temperatures are at or above the optimum for economic output, increases in temperature consistently led to decreases in GDP. When temperatures increase in the United States, India, or China, GDP declines in every case. GDP declines because crops fail, conditions become too hot to work outside, brains become addled, and, directly or indirectly, violence breaks out.

An obvious question emerges in response to these results: Do humans just need time to adapt through new behaviors, cultural practices, or technologies? Perhaps the decline of GDP with increases in temperature is just the jolt of the new, and if countries have a chance to adjust working hours or use new technologies, productivity will recover. Hsiang, Burke, and Miguel had two ways to consider this. First, they compared responses of GDP in countries between 1960 and 1989 to those between 1990 and 2010. Their intuition was that because global temperatures have been increasing since 1960 (indeed, since well before), the warming in those first twenty-nine years might have led countries to adapt to their newer (ever-warmer) conditions so that the negative effects of warming, above the optimum for economic output, in the subsequent twenty years would be less pronounced. But they found no evidence of such adaptation. Warming of countries to temperatures above the human optimum was just as much of a problem between 1960 and 1989 as it was between 1990 and 2010. This doesn't mean humans couldn't learn to adapt. It does, however, mean that even given twenty years to do so, they didn't.[8]

The other way to address the question of adaptation is to consider countries' relative affluence. It has been hypothesized that affluent countries might buffer the effects of climate through their wealth. If nothing else, more work in affluent countries is done indoors, so the direct effects of temperature on bodies might be less. More-affluent countries might also be able to use technology to buffer the effects of drought due to extreme heat and declining precipitation by using desalination plants, for example. Hsiang, Burke, and Miguel, however, found no correlation between more affluence and less decline in GDP. Just like poorer countries, affluent countries suffered too. The overall story, then, is remarkably simple. Increasing temperatures above the temperatures associated with the optimum human niche lead to rising violence, decreases in GDP, and, circling back to Xu's work, a reduced probability of being able to sustain large populations.

KNOWING THE NICHE associated with contemporary, high-density, human populations allows us to consider where that niche will shift in the future, what humans will need to do in order to home in on the conditions needed to thrive, especially at high densities. That is, ecologists can track the ways humans will need to move much as they have done with birds or plants. When Xu and his colleagues did this, they found that the places well suited to human success and survival will shrink in the future and move north in the Northern Hemisphere and slightly more idiosyncratically in the Southern Hemisphere. The desideratum moves into Canada in North America and Scandinavia and northern Russia in Europe and Asia. Meanwhile, northern sub-Saharan Africa, the entirety of the Amazon Basin, and roughly half of tropical Asia move farther from the optimum of the human niche by 2080 under one of the climate change scenarios in which we dramatically reduce emissions, RCP4.5, and outside the human niche by 2080 under the business-as-usual climate scenario, RCP8.5. Unfortunately, it is in these very same regions that human populations are expected to grow most quickly in the coming decades. As a result, many people are predicted to be living outside the human niche conditions by 2080. Under what many now regard as the best-case scenario with regard to our global attempts to rein in greenhouse gas emissions, RCP4.5 (introduced in Chapter 4), 1.5 billion people will find themselves living outside the human niche in six decades. Under RCP8.5, the business-as-usual scenario, 3.5 billion people will find themselves living outside the human niche in six decades.

Conservation biologists have thought a great deal about how to help the species that must move in light of climate change in their search for new homes. The approach of creating corridors and conserving as much habitat as possible is not perfect, but it is nonetheless an approach that has some momentum and will help many thousands and perhaps hundreds of thousands of species.

We will need to find a way to help hundreds of millions or even billions of people to find their new homes. We need an ambitious global plan, a plan that recognizes not only the large numbers of people who will need to move, but also another aspect of geography. To date, the vast majority of the greenhouse gases that have contributed to climate change have been produced by people and industries in the United States and Europe. But the consequences of these greenhouse gases on climate change, and in turn, on people, will be disproportionately borne by people in regions currently near the extremes of the agricultural niche, people who have played essentially no role in creating the greenhouse gas emissions. The obligation to help in the homing of millions of families, to create corridors for their survival and success, falls most heavily on the countries that engendered the crisis we are facing.

Meanwhile, there is also one more hope in the regions likely to find themselves furthest from the niche for agriculture-dependent peoples. If you revisit Figure 5.1, you will see that while the main human niche is very narrow, confined to the same circle of temperature and precipitation that was being most heavily used six thousand years ago, there is also another climate space included in the modern human niche, a climate space that is very hot and very wet. In their paper, Chi Xu and his colleagues note that this space largely corresponds to the monsoonal parts of tropical India. Xu and his colleagues do not attempt to explain this particular extension of the human niche. But one possibility is that Indians have found cultural approaches to dealing with bodily heat and have also found agricultural approaches to dealing with heat's effects on sustenance. In their studies, Xu and his colleagues found that not only is the climatic niche of India hotter and wetter than anything that has existed before, so too is the climatic niche of Indian crops and domestic animals. Here is hope. The implicit suggestion in this example is that we need to, very soon, identify all of those places where peoples have found ways to live in a human niche outside the ancient human niche and learn from the successes of those places and modify them. The more we can broaden the human niche, the less suffering the future will entail.

But we must remember that while India may offer some answers for regions that will have climates in the future like the climate in India today, the increases in temperatures in India itself are expected to engender conditions unlike those ever experienced by humans. And not just India. Much of the world's population is predicted to be living in conditions warmer than those found in the hottest parts of India by the year 2080, especially under the business-as-usual scenario, but even under the most optimistic scenarios.[9]

CHAPTER 6

The Intelligence of Crows

ON THEIR OWN, THE CHANGES IN AVERAGE TEMPERATURES EXPECTED in the coming years are enough to lead to crippling effects on people, cultures, countries, and millions of wild species. The world will suffer the deadly heat of our actions and inactions. Unfortunately, these *average* changes will not occur in isolation. They will be coupled with increases from one year to the next in the variability of precipitation and temperature.[1] "Variability" sounds both vague and harmless. It is neither; it is, instead, one of nature's greatest dangers, an elemental threat. Variability is to be feared. Variability needs to be planned for.

Many wild, nonhuman animal species can respond to changes in average conditions by moving along corridors or through the air to places that are more suitable (which is to say, they can home). Scientists have also documented some cases in which species have undergone rapid evolutionary responses to recent changes in average conditions. For example, ants living in hot parts of Cleveland have been shown to have evolved a higher tolerance for hot temperatures than their country cousins.[2] Nature selected against the lineages of ants unable to deal with high temperatures. Natural selection can help species deal with new conditions, as it has done for billions of years, one birth or death at a time.

But rapid adaptive change in simple attributes of organisms, such as their ability to withstand heat, is most useful to species when the new conditions in one year are predictive of those they will encounter in the next year. Adaptive changes work well, for example, if the conditions of the future are warm, warmer, and then warmer still, but not as well if the future's conditions are variable, if they fluctuate from warmer, to very cold, to warmer than before, and so forth. Yet in many regions we are already seeing the latter pattern, long-term warming trends punctuated by unusual extremes. Parts of Texas have seen "unprecedented" heat, drought, and fires followed by record cold. Australia has suffered from record droughts, followed by rains that have flooded cities. In the future, such variability will become even more common and extreme.

The problem for species that might adapt to variable conditions is that they are pushed to one extreme one year and another the next. On the Galapagos island of Daphne Major, for instance, an El Niño event in 1982 caused prolonged rainfall, which led one of the plant species on which some of Darwin's finches depend, a plant with big seeds, to become rare. In that year, individuals of the medium ground finch (*Geospiza fortis*) with smaller beaks were more successful than individuals of the same species with larger beaks.[3] The next year, in 1983, more of the finches had smaller beaks. The finches had evolved. And because the big-seeded plant species continued to be rare, the smaller-beaked finches continued to thrive. But had the big-seeded plant species bounced back in 1984, when the El Niño event was over, the story would have been very different. The finches would have had exactly the wrong equipment for their new condition. Big-beaked medium ground finches would have been favored. Natural selection can push a species back and forth like this for a long time, but eventually the push and pull becomes too much. Eventually, a bad year, yet another "different" year, leads not to adaptation but extinction.

So what kinds of adaptations might occur in light of variable conditions, and what sorts of species might have those adaptations? For which species is variability itself part of their niche? And, importantly, can we learn to be like those species? For animals, a law provides the answer to these questions. It is the law of cognitive buffering, whose basic idea is that animals with big brains are able to use their intelligence in inventive ways so as to find food even when food is scarce and maybe also warmth when it is cold and shade when it is hot. They buffer bad conditions with big brains. Superficially this would appear to be a law that bodes well for us humans. We have very big brains relative to our bodies, big enough that our heads nod with their own weight when we are exhausted. What's more, those big brains are thought to have evolved, in part, to help cope with variable climates. But whether our big brains will help us in the future is going to depend on just how we use them, whether we and our institutions are more like a crow or more like the dusky seaside sparrow.

To explain the crow and the sparrow, I need to introduce two ways that birds use their brains to confront their daily challenges. Some birds have what I'll call inventive intelligence, the intelligence necessary to alter their behavior and, in doing so, invent new solutions to new problems and new conditions. Inventive intelligence leads birds to both think their way through new challenges and

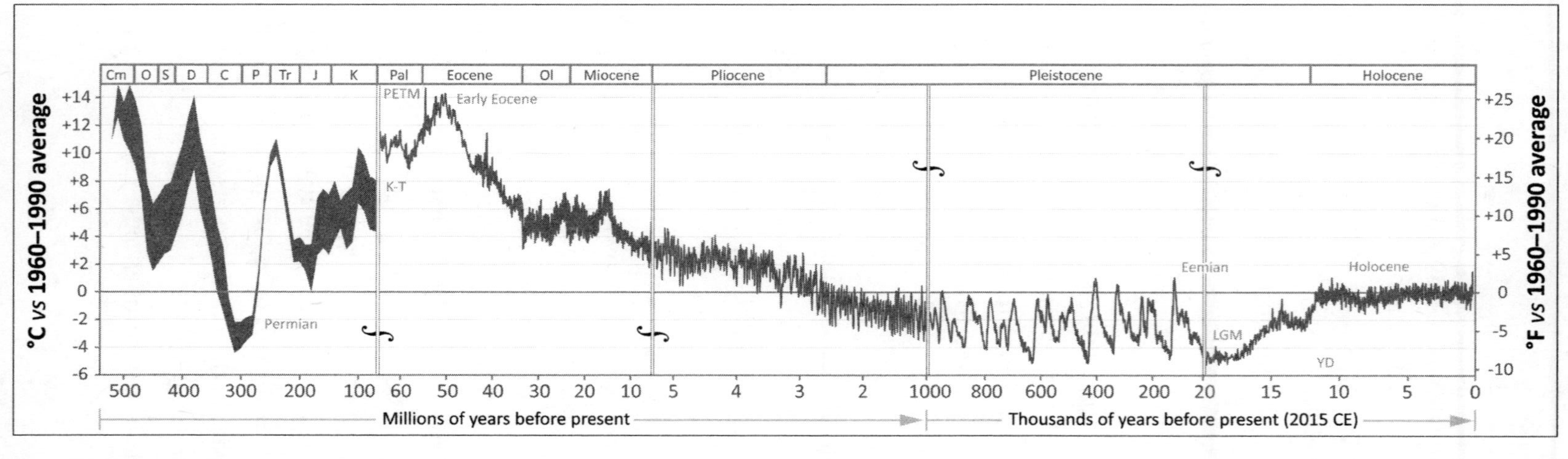

Figure 6.1. The history of climate change as reconstructed based on proxies for climate extracted from ice cores and other sources. Over Earth's history, climate has changed again and again. However, three things make the current change relatively unique. The first is its speed. The warming occurring now is occurring faster than any warming that has occurred for millions of years. The second is the magnitude of warming. Warming like that projected for the next century last occurred during the Eocene, more than forty million years ago. Third, the climate that humans have experienced since the dawn of agriculture has been extraordinarily stable, as shown on the far right of the figure. Our cultures and institutions evolved in the context of this stability. Future climates will be marked not by stability but, instead, by variability, whether among seasons, among years, or among decades. This figure is a modification by Neil McCoy of a figure generated by Robert Rhode based on data from Lisiecki, Lorraine E., and Maureen E. Raymo, "A Pliocene-Pleistocene Stack of 57 Globally Distributed Benthic d18O Records," *Paleoceanography* 20 (January 2005): PA1003.

to learn to repeat the solutions. Inventive intelligence helps birds both remember where they have stored food and use that stored food when it is most necessary. Inventive intelligence also helps birds find novel ways to access food. New Caledonian crows use tools of different sorts to access food they can't otherwise reach. They fashion those tools. In the laboratory, when presented with a food it couldn't reach with a straight wire, a New Caledonian crow named Betty bent the wire into a hook. In the wild, different populations of New Caledonian crows use different tools to do different things.[4] The crows learn and invent. As John Marzluff and Tony Angell note in their lovely and fascinating book *Gifts of the Crow*,[5] inventively intelligent birds invent in ways that neither the smartest dogs nor the smartest human toddlers can. They meet novel situations with novel behaviors. They are, as the evolutionary biologist Ernst Mayr put it in describing ancient humans, specialized at despecialization, specialized at doing different things in different times and places.[6]

Inventive intelligence is not the only way, however, that birds negotiate their daily challenges. Birds can also have the sort of know-how associated with specialization. It is the know-how of being able to do a set of narrow tasks very well. These species, as the writer Annie Dillard put it, "grasp their one necessity" and do "not let it go."[7] Pigeons find their way home even taken thousands of miles from their roost. Vultures find dead animals from miles away. Quails burst startlingly forth, together, in response to danger. Cormorants know when and how to dry their blue-black wings. These examples of know-how are not inventive; often they do not even involve the brain so much as the rest of the nervous system spread throughout the body and connected in the most ancient, involuntary part of the brain. Thus, one might call these forms of know-how involuntary or autonomic.

One bird species with a remarkable autonomic know-how was the dusky seaside sparrow. Dusky seaside sparrows lived in marshlands around Merritt Island and along the nearby St. John's River in Florida. For thousands of years, they made perfect use of the marshland's grass stems for nesting areas and of the insects among the grass stems for food. They had the know-how necessary to know where to be; they had a behavioral tendency to fly, eat, mate, and live only on and around Merritt Island and along the St. John's River, where things were perfectly suited to their way of life. They lived nowhere else. In general, they had the kind of know-how necessary to do their one thing, living like a dusky seaside sparrow, nearly perfectly. In this, they were like thousands of bird species.

Birds with inventive intelligence are predicted to thrive in a variable future. Birds with specialized autonomic know-how are predicted to suffer. More specifically, they will suffer the consequences of not letting go of a way of life that is disappearing. Jumping ahead in this chapter a little, it isn't an enormous leap to also imagine that human institutions and societies with inventive intelligence might thrive, while those with specialized know-how might suffer. But we'll come back to us; let's stick, for now, with the birds.

Amazingly, scientists more or less agree on how to measure inventive intelligence, at least in birds. Birds with bigger brains relative to their body size do more inventive things. Daniel Sol, a researcher at the Centre for Ecological Research and Applied Forestries in Catalonia, Spain, is a leading scholar when it comes to thinking in birds. He has studied intelligence in birds for two decades. In 2005, Sol documented that big-brained birds are generally more likely to engage in novel feeding behaviors, whether trying out a new way of eating a familiar food or trying an unfamiliar food.[8] There are, of course, exceptions. Some big-brained birds aren't terribly flexible, and some small-brained birds find ways to be creative. But broadly speaking, the pattern holds.

Big-brained birds include crows, but also ravens, jays, and other species of the family Corvidae, as well as parrots, hornbills, owls, and woodpeckers. Then, of course, within each group of birds, some are brainier than are others. House sparrows can think circles around other species of sparrows. The subset of these species with the very biggest brains are sometimes called feathered apes. For good reason. The average human brain is about 1.9 percent of human body mass. That of a raven, as reported by Marzluff and Angell, is 1.4 percent of its body mass—slightly smaller, but only just slightly. Meanwhile, the brain of a New Caledonian crow is 2.7 percent of its body mass. Mammal brains and bird brains are sufficiently different that such comparisons shouldn't be taken too seriously. Nonetheless, suffice it to say that crows are pretty brainy, akin to "feathered apes," though it is probably just as accurate to refer to apes as "flightless crows."

Birds with autonomic know-how are more diverse, reflective of the many different kinds of ways to specialize; what they share, other than their specialization, is that they all tend to have small brains relative to their body size.

What the tentative agreement with regard to which birds abound in inventive intelligence allows is a consideration across many species of whether inventive intelligence is one of the things that helps species cope with variability in conditions, specifically variability in climate, whether from year to year or even among seasons. Scientists can test whether birds are more likely to evolve inventive intelligence in regions or biomes with variable climates. They can also test whether intelligent birds tend to move into novel, human biomes with variable conditions when they appear.

This seems likely to be another case in which disagreement would reign. But again, there is broad agreement.

SOME OF THE recent research elucidating the law of cognitive buffering was led by my friend and collaborator Carlos Botero. It was through Carlos that I learned about the law in the first place. Carlos grew up in Colombia, stumbling over impediments while looking up at birds. Birds led Carlos to Cornell University in New York and then on to Washington University in St. Louis, Missouri, where he is now an assistant professor. Carlos became fascinated by bird behavior, initially focusing on the song-making abilities of male tropical mockingbirds. He discovered that mockingbirds produce more inventive, elaborate songs in more variable environments; it was Carlos's work on the songs of mockingbirds that led to his broader interest in bird brains, bird intelligence, and the question of which bird species are likely to thrive in the variable future.

Carlos and his friends and colleagues have studied several kinds of variability faced by birds. One kind relates to the differences in temperature and precipitation within years, which is to say, seasons. Such variability is predictable (it happens every year) and yet still a challenge. Where birds must deal with seasons, Carlos and others have found that, yes, they are more likely to have bigger brains. This is true if one compares different groups of birds—for example, if one compares corvids, such as ravens, crows, and magpies, to flamingos. Seasonality also favors big-brained species within particular groups of birds—for example, among owls. Owl species that live in seasonal environments tend to be especially brainy.[9] Those bigger brains help them find food where it is otherwise scarce. Other researchers have shown that the same is true in comparisons among species of parrots.[10] These patterns can even be seen within species; a recent study by Gigi Wagnon and Charles Brown at the University of Tulsa found that smaller-brained cliff swallows are more likely to die in extreme cold snaps than bigger-brained cliff swallows.[11] Conversely, birds that live in seasonal environments but migrate, and hence escape the consequences of seasonality, not only don't have big brains but, instead, tend to have especially small brains. They are all wing.[12]

As a caveat, a kind of small subplot to the story, a number of researchers, including Carlos Botero and his collaborator Trevor Fristoe, as well as Daniel Sol and his collaborators, have found that big-brained birds are not the only species that do well with seasonal variability. So too do the subset of small-brained bird species that happen to have lifestyles attuned to the specific kind of variability with which they are confronted.[13] For example, if the variability is due to winters, bird species with tiny brains—peanuts rather than walnuts, and often just half a peanut—can do well if they happen to also be very large and to have big guts that they rely on to ferment foods. Such birds have the specialized know-how necessary for the specific details of the variability they face. And so, for instance, in the far north, where summers are warm and winters cold, ravens thrive, crows thrive, and owls thrive, but so too, as Carlos points out, do grain-, pine needle–, root-, and stem-eating small-brained grouse species and pheasants.

But in some ways, the kind of variability associated with seasons is the easy kind. Even if it is a shock to the system each time it comes—the shock of the first snow, the first spring rain, or the first summer heat—it is an expected shock. Spring. Summer. Fall. Winter. Spring. Summer. Fall. Winter. Another kind of variability relates not to differences among seasons but, instead, to differences among years. Such variability is more difficult to deal with because it doesn't have a pattern. A bird can't anticipate a dry year.

And it is this unpredictable variability, the unpredictable year-to-year variability in temperature and precipitation, that is expected to increase in the future. Like seasonal places, places where conditions are different from one year to the next also tend to favor inventively intelligent birds.

Inventive intelligence is likely to often help birds by allowing them to find food, even when their ordinary foods are scarce. It helps them diversify the species on which they can prey. I think about the value of the inventive intelligence of birds in the context of one of my own recent experiences watching crows. I spend part of each year working at the University of Copenhagen. The last time I was in Copenhagen, I often observed a group of hooded crows on my bike ride to work. These crows, close kin to American crows, gathered at the beach on the road heading north along the coast out of the city. I passed this same group every day. As a result, I could keep tabs on what they were eating. In the late summer they ate human food, bits and pieces of rye bread, French fries, and potato chips along with sips of, this being Denmark, Carlsberg beer. But as temperatures cooled in August, beachgoers became sparse and less trash was available. The crows switched to walnuts gathered from a nearby tree; they could be seen, all day, dropping walnuts again and again onto the cement of the sidewalk to crack open the outer husk and then dropping them again to crack open the shell. When the walnuts were gone, the crows dropped apples. When the apples were gone, they dropped mussels. Recently, as I biked by, I saw them dropping snails. Despite being at an edge of the city where the bounty of wild nature was not very apparent, the crows innovated in order to get by. Theirs is precisely the sort of innovation that Daniel Sol found tended to be associated with having a big brain. The crows appeared to use their big brains to find, choose, and access new foods, in ways that were useful in coping with monthly variation within the city but that would also be useful in coping with year-to-year variation. Everyone who watches crows with any patience ends up with their own examples of inventive culinary practices. Nor is it just the crows. Blue tits in one neighborhood in Great Britain are reported to have learned to peck through the aluminum caps on milk bottles left on porches to get to the cream. In *The Beak of the Finch*, Jonathan Weiner reports that this practice, once it was invented, spread through the neighborhood, bird to bird and porch to porch.[14] When other birds might have been suffering, the inventive blue tits were living on life's cream.

But eating different foods in different seasons and inventing new ways to get to new foods are just part of how inventively intelligent birds cope with variability. They also store food. For example, Clark's nutcrackers are able to store piñon pine nuts by burying them. Individual nutcrackers are then able to use their big brains to remember exactly where they buried individual pine nuts. Their brains help them know when to store the nuts. They help them know where to store the nuts. And they help them know where to dig up the nuts they have stored. An individual Clark's nutcracker can remember the location of thousands of individual nuts even ten months after they were buried. One could argue (though I won't) about whether what is necessary to remember the location of the nuts is truly inventive intelligence or, instead, a unique form of know-how. But the component of this intelligence that is definitely inventive is the ability to retrieve nuts at some times and not at others and to choose which nuts to retrieve first and which to save. The birds not only store food, they also carefully ration it. For example, scrub jays, as Marzluff and Angell point out, "recover perishable worms sooner than nonperishable seeds,"[15] which is to say, they employ a sort of avian "best by" label. Nor are these by any stretch the limits of the abilities of inventively intelligent birds. Marzluff and Angell note that both ravens and scrub jays will rehide food they have stored if they notice that other birds—potential pilferers—watched where they hid the food in the first place.

If this idea about the buffering effects of intelligence is right, one can make some other predictions. If being able to find diverse solutions to problems helps bird species when conditions are variable, then the populations of big-brained bird species should vary less, from year to year, in variable climates than do those of small-brained birds. Carlos Botero, along with Trevor Fristoe, has shown this to be the case. In good years, small-brained bird species wax; in bad years they wane. The populations of big-brained bird species are steady—buffered.[16] One might also predict that big-brained bird species should be more likely to thrive when introduced by humans into variable climates. They are.[17] And one should expect big-brained bird species to be more likely to thrive around humans, in cities, where conditions are unpredictable, both from one patch of the city to another and from one time to another. The evolutionary biologist Ferran Sayol, while working with his adviser Daniel Sol and another mentor, Alex Pigot, showed that they are.[18] The other species that do well in cities are small-brained species with an unusual specialization: they breed often. Species that breed often survive in cities by making many babies and "hoping" that some will end up in the right place or time to thrive.

With regard to big-brained species in cities, think corvids. Hooded crows in Copenhagen. Pied crows in Accra, Ghana. Jungle crows in Singapore. Fish crows in Raleigh, North Carolina. As the poet Mary Oliver put it, "At the edges of highways / they pick at limp things," the "deep muscle" of the living world in cities.[19] In her book *Crow Planet*, Lyanda Lynn Haupt has gone so far as to argue that there are more crows and other corvids now than at any time in Earth's history.[20] Maybe. Maybe not. Certainly though, a subset of corvids is succeeding alongside us.

But it isn't just corvids that use their brains to succeed in cities. So too do owls. So too do even a subset of parrots. The rise of smart birds around us is a measure of just how unpredictable we have made the world. Crows and other intelligent birds are indicators of the conditions most species cannot tolerate, unpredictable conditions. On January 12, 1855, Henry David Thoreau wrote in his journal that the crow's caw "mingles with the slight murmur of the village, the sound of children at play, as one stream empties gently into another, and the wild and tame are one."[21] To Thoreau, the crow spoke for itself, but also for him. But it is probably more accurate to say that the presence and abundance of crows spoke not for him or for us but, instead, about us.

WHICH BIRD SPECIES suffer with increases in variability? They tend to be the birds with kinds of specialized know-how that no longer match up with whatever the new conditions might be. Such birds try to survive tough times by continuing to do what they have always done. They cling to the old ways no matter the cost. Such was the case, for example, with the dusky seaside sparrow.

I already mentioned that the dusky seaside sparrow lived on and around Merritt Island, at the end of the Canaveral Peninsula. There, and nearby along the St. John's River, the sparrows specialized on high, relatively dry marshes in which they evolved over two hundred thousand years. The long-term stability of the marshes meant that the birds did not need the kind of intelligence necessary to respond to new conditions.

What I did not mention is that Merritt Island also happens to be where the National Aeronautics and Space Administration

(NASA) decided to locate the John F. Kennedy Space Center. NASA chose Merritt Island as the place from which it would send rockets into space, rockets from which humanity would look back on Earth. It was from the Kennedy Space Center, for example, that astronaut Mike Collins launched into space as part of the Apollo 11 mission. It was thinking about that mission that would prompt Collin to later say during an interview for a documentary, "The overriding sensation I got looking at the Earth was, my god that little thing is so fragile out there."[22]

Both prior to and after NASA's decision to make Merritt Island the focal point for its space program, the umbilical cord between Earth and space, the human use of Merritt Island had come to include efforts to control the island so as to make its conditions more suitable and consistent for humans. The first effort at control was the use of the pesticide DDT. DDT was sprayed over the island in an attempt to control the island's mosquitoes. That spraying had two effects. It killed large numbers of the insects the dusky seaside sparrows ate. And it inadvertently but entirely predictably favored the evolution of mosquitoes (and presumably some other insect species) resistant to DDT. The reduction in the total living mass of insects appears to have led to major declines in the numbers of dusky seaside sparrows. Spraying started in the 1940s. By 1957 the sparrow's population had declined by 70 percent. The sparrows did not have the inventive intelligence necessary to find something else. Meanwhile, once the mosquitoes had become resistant to the DDT, new measures were employed to control them. These new measures were ambitious in no small part because by then the island was inhabited by the many people working at the space center. The marshes were ditched or impounded. They were subject to Noachian floods or they were drained and dried or one and then the other. The habitat left that matched the sparrows' key necessities was thus further reduced and made into ever-smaller island-like patches. The best of those patches shrank further when a highway was built connecting Walt Disney World to the space center. The highway led to the construction of houses, which led to more flooding and more drainage. In 1972, a survey found 110 male dusky seaside sparrows, and hence there were about 200 birds total. A survey in 1973 found 54 males, and hence about 100 birds. A survey in 1978 found 23 males, and so 50 birds. Then there were just 4 males and no females. The last dusky seaside sparrow died, in a cage, in 1987. It had been brought in from the wild so that it might be bred with another sparrow species to ensure, in some altered form, its persistence. "Ironically," as one report pointed out, that caged male, which still sang, "was housed at Disney World," the "world" that, in some form, had replaced its own.[23]

The dusky seaside sparrow was a small bird with a small presence in the universe. But since its passing, much has been said about its loveliness. It has been the subject of fiction, poetry, and numerous scientific treatises. As the writer Barry Lopez put it, "As with so many things, deep appreciation and a sense of loss arrived simultaneously."[24] Ultimately, the dusky seaside sparrow was a victim of the intersection of its own specialization and the forces of development, technology (rockets), politics (the space race), and pleasure (Disney World). It could no more have anticipated these forces, with its specialized autonomous intelligence, than it could have anticipated a meteor.

The dusky seaside sparrow is not alone in suffering the consequences of the variability that we leave, everywhere we go, in our wake. Insect-eating birds have declined worldwide, in no small part because of declines in insect populations. More generally, though, the many birds with specialized, wondrous ways, specialized ways

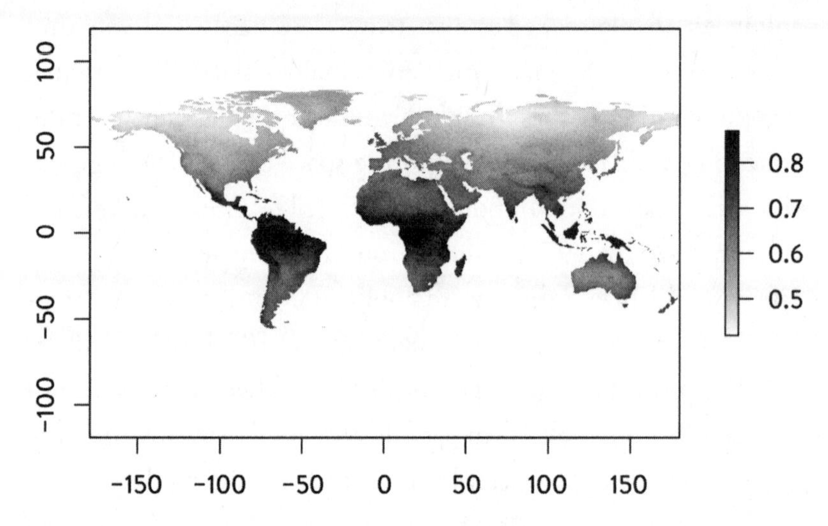

Figure 6.2. The historic predictability (from one year to the next) in temperature. Dark-shaded regions have very invariant (and hence predictable) temperatures. An animal that evolves to respond to the temperature of one year probably has the traits necessary to deal with the subsequent year. In lighter-shaded regions, on the other hand, the temperature in one year is not necessarily predictive of that in the next. The most inventively intelligent birds are disproportionately likely to thrive in the lightest-shaded, most unpredictable regions, whether that be central Australia, North Africa, temperate Asia, or North America. Figure produced by Carlos Botero.

of being smart in their world through living their one necessary way well, are declining.[25] Hundreds of bird species are now extinct due to the ways humans have altered the world during the great acceleration.

When a child blocks a stream with sticks, the child has momentarily controlled the stream. The flow stops. Everything is controlled and dry. But then the water pours over the small dam. What was a stream rushes down its old channel and is, for a moment, a river. Often, in controlling things, we introduce variability. In our efforts to control the world and make it less variable, we have made it more variable for other species in the short term. In the long term, we have made it more variable for ourselves because our many small decisions—the cars we drive, the trips we take, the foods we eat, the numbers of children we have—have added up to flood greenhouse gases into the atmosphere and alter the climate. We should ask ourselves how we will respond to this variability, whether we are more like a crow or a dusky seaside sparrow.

As we ponder this question, it is useful to first know that recent studies have shown that mammals, like us, are not immune from the law of cognitive buffering. For example, one study considered which mammals are most likely to survive after having been introduced to regions with conditions different from those in which they evolved—conditions that are, to them, suddenly novel. The study found that animals with big brains were far more likely to survive.[26] As a result, the mammals we have spread with us around the world and also those we are unconsciously willing forward through our attempts at control are the ones with inventive intelligence.

Within primates, the role of intelligence has been a special focus, in no small part in attempts to explain ourselves. We ask, "Who am I?," and we turn to monkeys and chimpanzees for answers. With primates, things are slightly more complex than they are with birds or with mammals more generally, but not *so* complex, so hear me out.

The first complexity is that with the exception of species of our own genus, primates have not really made it out of Earth's most predictable climates. (By that same token, it is likely that many primate species will suffer disproportionately due to climate change.) Of course, we are the primate with the biggest brain, and we live in the most unpredictable climates. But considering ourselves makes things challenging. We are too close to our own story to see ourselves clearly. Studying the key factors that influenced the evolution of big human brains is a little bit like trying to look at the back of your head in a mirror; you can do it, but the perspective is never quite right. It is easier to focus on the nonhuman primates.

In nonhuman African primates, as in lineages of big-brained birds, brains get so relatively big that they can be very energetically costly. In this context, there are two competing explanations of how climate variability and unpredictability might be related to brain size and hence to inventive intelligence. One is that, just as with birds, primate brains should get bigger when the climate is unpredictable: big brains and their cognitive abilities buffer the effects of hard times. The other explanation is that if unpredictable climates lead to less food, primate brains should actually get smaller relative to body size, there not being enough food left over to invest in brains. Instead, primates should evolve small brains and high fecundity.

One way to consider these two alternatives together is to consider not just brain size but also a more direct measure of what brains allow primates to do, the extent to which primates are able to eat the same calories and nutrients per day independent of differences in climate and habitat variability. Here the idea is that an inventively intelligent primate finds ways to eat enough, even in hard times. An inventively intelligent primate, in other words, can do what the hooded crows I biked past in Copenhagen were doing. It can eat French fries when there are French fries and nuts when there are nuts. When researchers recently tested this, they found a situation slightly different from that for birds and yet similar. Primate brain size actually tends on average to be smaller in variable climates than in less variable climates and hence to require fewer calories. This observation is in line with the idea that big brains are energy expensive and are sometimes, when conditions are hard, just not worth all the fuss. However, the primates that are able to eat the same number of calories, regardless of climatic variability, are those with bigger brains.[27]

Put another way, you can be a primate in variable environments either by having a small, low-cost brain, and often, also, a small body, or by having a big brain and using it to find new ways to get enough calories. The primates most able to do the latter include guenons, baboons, and chimpanzees. Chimpanzees, to take the example for which we have the most data, are able to eat similar things whether they live in wet forests or savannas. They do so by remembering where fruiting trees are and when they will be in fruit, but also by using their brains to make tools that allow them to eat food that would otherwise not be available, food such as algae, honey, insects, or even meat. Indeed, recent work by my collaborator Ammie Kalan, a researcher at the Max Planck Institute for Evolutionary Anthropology, shows that chimpanzees are most likely to use tools at sites where conditions are unpredictable.[28] At a site called Fongoli, in Senegal, for instance, chimpanzees have found a way to find meat even where their preferred prey species aren't present. They make spears and thrust them into holes where bush babies are sleeping.

Building on the sort of inventiveness and tool use that chimpanzees employed, human brain size evolved to be ever bigger in unpredictable conditions and times. With those big brains, humans buffered variation across those conditions and times. This doesn't mean that such climates are the whole story of the origin of our brains (they almost certainly are not); instead, it means that our own history seems to match those of many other species. We chose the road well traveled.

THE MOST OBVIOUS implication of the cognitive buffering law for the future is with regard to which species will thrive in the changing world. Consistent warming conditions will tend to favor species able to tolerate such conditions, species with the right climate niches. Similarly, warm, wet conditions will favor species with niches suited to warm, wet conditions. Warm, dry conditions will favor species with niches suited to warm, dry conditions. Very cold conditions will favor species with niches suited to very cold conditions, or they would if very cold conditions were to exist in the near future. For the most part, they won't. But variable conditions may favor quite different species, species with niches that include variable climates; the world may become ever more a place of crows and rats and, by that same token, ever less a world of dusky seaside sparrows and the thousands of species like them.

The other implication of this law has to do not with species but, instead, with our own societies. As Marzluff and Angell note, "The writings of early Scandinavians celebrated ravens as useful informants,"[29] and the first peoples of the Pacific Northwest of North America saw them as "motivational forces." Similar sentiments are voiced by the Indigenous peoples of the far north. Perhaps insights and motivation from these clever birds can still inform us today. But what might that motivation look like? How do we live like a crow?

Once, when we all lived as hunter-gatherers in small groups, the same sort of intelligence that benefits the crow benefited us. This was especially true in the variable, unpredictable far north and deserts of North America and Australia. In such places and times, humans used their own crow-like inventiveness to deal with new conditions. In fact, in many of the regions in which inventive intelligence benefited crows, it also benefited humans—so much so, that humans and crows often found their lives to echo one another. In what is now the southwestern United States, Indigenous people gathered the same piñon pine seeds that the Clark's nutcracker gathered. They too stored them. They not only did as the crow did, they competed for the same food to be stored, for leaner times, in much the same way.

However, most of us no longer live in the old ways. We are no longer in charge of the means of the production on which we rely. We do not grow our own food. We do not build our homes. We do not construct the transportation systems, waste-treatment systems, or education systems on which we depend, not individually anyway. Most of us couldn't do these things if our life depended on it, not just because we lack the ability but also because we now live in cities. In cities, we rely on systems to carry out these tasks, systems that are run by humans and yet have, within them, rules that engender a kind of intelligence apart from that imparted by individual human brains. If we are to think about our collective ability to respond to variability in the future, we need to think not about our own brains but, instead, about the brain-like workings of our public and private institutions.

Just as with animals, we can imagine that institutions might have different kinds of intelligence. Many, perhaps most, institutions are focused on being able to do one thing perfectly, or if not perfectly, at least pretty well. They have specialized autonomic know-how. Universities have a tendency toward this model, as do governments. If they are effective, such institutions are effective in light of the average conditions over the last decades, and sometimes longer. Or as my colleague at North Carolina State University, Branda Nowell, an expert in how institutions respond to risk, put it, "We have large public bureaucracies that have structurally and culturally evolved over time in a myriad of ways to adapt to their dominant operating environment." They are as specialized on those "dominant operating environments" as the dusky seaside sparrow was on its world of salt spray and grass. In such institutions, one often hears a kind of patois of stability and specialization, a patois that emphasizes the past. People say, "This is the way we've always done it," which is code for "This is what's always worked." Sometimes what worked in the past is not a specific solution but an approach to solving a problem. But even then, the use of such an approach assumes the conditions are similar enough for the approach to be meaningful. As Nowell put it, in a changing world, often "the relationship between past action and past outcomes has potentially limited relevance to the current situation."[30] The ancient rules of cause and effect must be replaced by new rules. Unfortunately, institutions with specialized autonomic know-how are very slow to implement new rules.

Other sorts of institutions may be more versatile. They might respond to changing conditions through inventive intelligence and reinvention. But to be honest, it is hard to come up with good examples of institutions with inventive intelligence. Maybe this isn't surprising. Our institutions, most of them anyway, arose in decades of relative stability. Our global economic system after the end of World War II was stable. But more importantly, we have become accustomed to climate stability. In the years since the evolution of *Homo erectus* and, later, *Homo sapiens*, with their big brains, Earth's climate has been more predictable than during most other periods of the last hundred million years. This has been especially true over the last ten thousand years (the Holocene in

Figure 6.1), the period during which agriculture, cities, and most features of our modern cultures emerged, the period during which the great acceleration occurred. We've been sheltered by a lucky stability for which we were unaware that we needed to be grateful. Put more succinctly, while our species evolved big brains in times of variability and unpredictability, our institutions evolved the kind of specialized know-how ideally suited to conditions like the ones we have long had, conditions that are, increasingly, no more.

One might expect that even in times of stability, many institutions would evolve so as to be ready for the change that might come, just as one might imagine that big-brained birds might, sometimes, evolve in invariant climates. One reason this might be rare is that the flexibility and awareness that inventive intelligence in institutions requires might come with costs, much as the big brains of primates themselves are costly. One of those costs is associated with deciding things anew each time rather than just doing what has always been done. "We already know how to solve this problem," a leader says; "we don't need to talk about it." This is the cost of time and paychecks, the cost of pausing to reflect and reconsider. In theory, this cost might be reduced if the system itself and its rules were designed to be responsive to changes. But even then, as highlighted by Branda Nowell, there is a cost in the vigilance required to detect when conditions have changed. What is more, as Nowell emphasizes, the sort of vigilance associated with the past may not be the sort of vigilance needed for the future.

The crow is ever aware. The crow knows when food is becoming short, when winter is becoming harsh. When these changes come, the crow invents. Big institutions are not, by their nature, aware of such changes. They must monitor for changes and be alert to them. They must be hypervigilant to events that have been rare. But the cost of doing so is that most years such rare events and changes don't happen. Most years, planning for such events entails costs that are made more apparent by the exigencies of quarterly reports. Until an oil company has a spill, its safety measures cost it greatly and return no benefit. Until a meltdown occurs, a nuclear power plant is wasting money training people in the ability to respond to an indication that a meltdown could occur. Or in the example on which Branda focused, until firefighters confront a fire freakishly bigger than has ever been seen, their readiness for such a fire is almost a kind of foolishness. As we look to the future, though, we know enough to be aware that variability will be increasing. As it does, the dangers of ignoring rare events and changes will become greater because they will become more common.

In the wake of the COVID-19 pandemic, it will be useful to study which sorts of institutions were more prepared for the risk that the disease represented. It will be useful because pandemics like the COVID-19 pandemic are predicted to become more common. Ecologists who study disease have known for decades that when disruption of natural ecosystems is combined with large-scale farming—or even just caging animals side by side—and globally connected human populations, new parasites will evolve. They've argued as much repeatedly. They've even pointed to the regions where the origins of such parasites are most common. They were like Babe Ruth pointing to the place where he would hit the ball out of the park. Ecologists pointed to where nature would hit a parasite into our collective society. But the big point is not that the risk of pandemics will increase; it is that the risk of many kinds of problems will increase—floods, droughts, heat waves, *and* pandemics—and so the extra cost of being inventively intelligent is ever less.

If we are to survive variability, our societies will need to be inventive. We can each pay attention to the signs of such inventiveness and to the changes necessary to achieve it, but we can also be conscious of those moments when we hear its absence, those moments when we hear someone, or even ourselves, say, "This is what we've always done" or "In this situation we've tended to . . ."

But there is also something else.

When crows use their inventive intelligence to deal with novel conditions, they do so by finding new ways to find foods and by eating novel foods. In essence, they diversify their diet so that even if a species they rely on becomes rare, some other species might be common. We too can take advantage of nature's diversity, whether in our farm fields or even on our bodies, in order to buffer our risk. We can even do so without being terribly inventive in our intelligence. Carlos Botero has shown that brood parasites, birds that lay their eggs in the nests of other bird species, can benefit from diversity even without being inventive. The subset of brood parasites that does well in variable climates does so by relying on more species of birds as hosts for their eggs,[31] so that if the population of one wanes, another might wax. They put their eggs in more than one type of nest, both so to speak and literally. We too can and should build such bet hedging into the ways we depend on other species. This won't necessarily work for all contexts. But, as I consider in Chapter 7, it can work in the context of agriculture. The Reverend Henry Ward Beecher once said, "If men had wings and bore black feathers, few of them would be clever enough to be crows."[32] Maybe not. And yet maybe we can still, nonetheless, buffer at least some of what might come.[33]

CHAPTER 7

Embracing Diversity to Balance Risk

THE GREAT AGRICULTURAL ACCOMPLISHMENT OF THE LAST CENTURY has been a matter not of sustainability or flavor or nutrition, but of quantity. Crop scientists set out to increase the number of calories produced per acre for human consumption. They succeeded. An acre of corn now produces more corn kernels, an acre of wheat more wheat grains, and an acre of soy more soybeans than would have been imaginable forty years ago, much less a hundred years ago. These increases in yields have kept many of the world's most important staples cheap and available and have led to decreases in hunger during the last decades.

These successes have been achieved through control. Through breeding and engineering, we have altered the genes of crops in ways that make those crops grow faster, especially when watered and fertilized. We have waded, as Annie Dillard has written, into the "wet nucleus of the cell" and inserted new genes that produce pesticides.[1] We have even inserted new genes that make plants resistant to herbicides (and then sprayed the fields with herbicide to kill back the nonresistant weeds with which they might otherwise compete). The defining feature of these manipulations has been that we have made our crops ever more part of our system of industrialization. Like components of an assembly line, they are controlled, and in light of that control they thrive. Many features of this system can be critiqued, though while bearing in mind that a far smaller proportion of the world's population is hungry today than a hundred years ago. Yet as we look toward the future, we see that this system faces a major challenge. We have built a food system that thrives when variability is minimized. But, as noted in Chapter 6, we have also altered Earth's climate in such a way as to make it much more variable and unpredictable. Herein lies a problem.

The industrial, technological approach to agriculture is well suited to helping solve some of the challenges of the future—for example, how to eke a few more calories per acre out of farms or how to produce more drought-tolerant crops. But it is not well suited to deal with variability, especially variability outside the scope of its control. The future's conditions will be a more rapidly moving target, particularly with regard to climate. The ideal crop for conditions this year is unlikely to be the ideal crop for conditions next year. What one hopes for in such situations is what ecologists call ecological stability. A stable natural ecosystem is one in which primary productivity, the amount of green life that grows in a particular area over a particular time period, does not vary much from one year to the next, even when climatic conditions do vary. A stable agricultural system is one in which year-to-year variation in yield is modest, even when climate is very variable. One approach to achieving such stability is to use technology to buffer environmental variation, to in essence keep conditions invariant. For example, you can water more when it is dry and less when it is wet, and with the aid of a mix of drones, weather stations, and artificial intelligence, you can do so with ever more precision. You can do so, that is, if you have the money. But this isn't the only approach.

The other approach to dealing with variability is inspired by nature. It is what crows would do if crows farmed. The approach, paradoxically, is to deal with climatic variability through varying which crops are planted so as to increase agricultural diversity—in other words, to fight one kind of variability by favoring another. The values of this approach first became clear in some fields of grass in Minnesota, fields in which an ecologist named David Tilman created a world in miniature in order to better understand the world at large.

AS A GRADUATE student, David Tilman figured out that he was a particular sort of ecologist, one who uses mathematical theory as a way of generating predictions and experiments in order to test those predictions. Initially, the experiments were relatively small.

One of the first experiments Tilman carried out aimed to understand how different species of algae coexist with each other. A pond might contain thirty species of photosynthetic algae, all needing basically the same nutrients, plus sunlight. Why didn't one of those species just win in the race to get the nutrients, causing the others to go extinct? One of the forefathers of ecology, G. Evelyn Hutchinson, called this mystery "the paradox of the plankton."[2] Tilman aimed to solve it, and to his own satisfaction, he did so. He showed, in a series of careful experiments, that algae species could coexist if they had different niches. In this case, their niches related to the resources they were most limited by (phosphorus and silica). Even though three species might all need phosphorus, silica, and sunlight, if one needed slightly more phosphorus, one slightly more silica, and one slightly more sunlight, they could coexist.[3] Insights derived from this experiment led Tilman to carry out more experiments on algae, experiments in which he tested other features of how algae coexist. On the basis of this work, Tilman, at the age of just twenty-six, was hired as an assistant professor at the University of Minnesota.

Though Tilman continued to work on algae in Minnesota, he also dallied with terrestrial organisms. For example, he studied the ants on cherry trees and the plants that grow around gopher holes at what was then called the Cedar Creek Natural History Area (now the Cedar Creek Ecosystem Science Reserve), some thirty miles outside of Minneapolis. While he was at Cedar Creek, Tilman decided to do another kind of experiment, something more lasting than a dalliance, an experiment to which his professional life would long be wed.

Tilman wanted to revisit some of the ideas he had tested in algae, but with land plants. In 1982, Tilman set up fifty-four plots in each of three abandoned agricultural fields (what ecologists call old fields) and a similar number of plots in a prairie. He identified and measured each individual plant in each plot. By chance, some plots were richer in their diversity, others poorer. Tilman then randomly divided the plots into one of seven different dietary regimes. The dietary treatments were different concentrations of fertilizer. At one extreme, some plots received no fertilizer. At the other extreme, some plots received as much fertilizer as is used in the highest intensity industrial agriculture. To make the project work, Tilman had to choose the fields, make the plots, feed the plots their allotted nutrients, and then, for years, study the results. Such work is hard. It tires the body the way farming does, but the fruits of the labor are all intellectual. A season of work in old fields yields insights, not plums.

Even in the first years of this hard work, Tilman made discoveries. He wrote several dozen papers about how plants do or do not coexist depending on the concentrations of different nutrients they receive. He wrote many other papers about how the plant community changed over time as a function of those nutrient concentrations. Some of the papers were lauded. Others were forgotten. But there was also something else. As years passed, Tilman was able to study the effects of his experiments on longer-term phenomena. In particular, he had the potential to test something called the diversity-stability hypothesis.

It had long been hypothesized that forests, grasslands, and other ecosystems that contain more kinds of species should be more stable, particularly in the face of major disturbances such as fires, floods, droughts, or plagues. The diversity-stability hypothesis predicted that more-diverse ecosystems should be less affected by such disasters. The plots that Tilman was studying differed from each other in the number of species they contained (their diversity), both because of the treatments they'd been given and because of chance differences in the history of the individual bits of land before Tilman began his experiment. Some plots were richer in species (which is to say, more diverse) and some were much poorer. The most diverse of the plots resembled natural grasslands. Some of their plants were tall; some short. Some had bunch roots; some had long straight roots. And they were many colored—a mosaic of browns, greens, and, as my friend Nick Haddad, who once worked in the plots, put it in an email, "visible, vivid flowers." The least diverse plots, which tended to also be those that were well fertilized, were most like intensively farmed crops. They tended to be quack grass or Kentucky bluegrass. Their plants were more uniform in height, uniform in the shapes of their leaves, uniform in their needs, and, also, uniformly green. This variation among plots had allowed Tilman to study the ways fertilization and other factors affected how many and what kinds of species were present in a plot. But, especially as months, seasons, and years passed, it would also allow him to test the diversity-stability hypothesis, to test whether his most diverse plots had varied less over time than the less diverse ones. Or at least it would allow him to test this hypothesis once disaster, whether it be fire, plague, flood, or drought, struck. He had to wait.

Of course, Tilman could have experimentally created some kind of disaster. He could have added parasites to his plots or set a fire. But he didn't have to experimentally re-create any of these horsemen of the apocalypse. The apocalypse came to him in the form of drought. Beginning in October 1987, five years after the experiment was established, the most severe drought in fifty years occurred in Minnesota. The drought was prolonged; it lasted two years. It was horrible. It was also just what Tilman needed. But he couldn't study the drought's effects immediately. Not only did he have to wait to see how the drought affected each of his plots, but he also had to follow those plots in the years after the drought to see how they recovered. The stability of an ecosystem over time is a function of its resistance. A resistant ecosystem does not change even in response to a disaster. It resists. The stability of an ecosystem over time is also a function of its resilience. A resilient ecosystem bounces back in response to disaster. Tilman could have studied the resistance of the plots as soon as 1989, but to study their resistance, resilience, and, ultimately, stability, he had to continue to wait.

Finally, by 1992, ten years after the experiment was set up and six years after the beginning of the drought, enough time had passed. And so Tilman began a collaboration with a professor visiting the University of Minnesota from the University of Montreal, John Downing, to study the resistance, resilience, and stability of each plot. Tilman and Downing decided to focus on the total amount of plant biomass, the living mass of life, produced in each plot in each year. They compared the change in biomass of the plots over time, which allowed them to measure each plot's resistance to the drought, its resilience after the drought, and the net consequence of its resistance and resilience, its stability.

What Tilman and Downing found was that plots with more species declined less in biomass after the drought.[4] During the drought, the biomass declined dramatically in the plots with few species. Their biomass was reduced by about 80 percent. Such plots were not resistant to the drought. Biomass still declined in the more diverse plots, but much less so, by about 50 percent. The diverse plots were thus comparatively more resistant. In addition, in the years that followed, the diverse plots were more likely to fully recover their biomass than were the low-diversity plots. They were more resilient. Because of their resistance and resilience, the diverse plots were, when considered across the years bracketing the drought, also more stable. Though such an effect of diversity on stability had long been hypothesized, it had not been experimentally documented in the wild. Here, then, was compelling evidence. More diverse grasslands are more stable. The diversity-stability hypothesis was looking more and more like a kind of law. Yet Tilman wanted to be sure. So, in 1995 he set up a new, even bigger, experiment.

The new experiment, dubbed the Big Biodiversity Experiment, or BigBio, focused on diversity. It aimed to even more directly

Figure 7.1. The photo at top shows some of David Tilman's larger biodiversity plots, part of the BigBio experiment. Notice the variation among plots in their shade, the amount of bare ground, and the height of the vegetation. The photo at bottom shows students weeding some of those same plots, one plant at a time. Photos courtesy of Jacob Miller.

characterize whether more diverse old fields might be more stable than other fields in the face of drought, parasites, and pests. The plots associated with this new experiment were even larger than those employed in the fertilizer project. And there were more of them. And all the plants in each plot were to be planted, by hand, from seed, after the existing vegetation was removed using road scrapers and plows. The resulting plots required nearly constant care. They needed to be measured, and especially during the summer, they needed to be weeded. The weeding was especially onerous, and so it was to prove the purview of what Nick Haddad remembers as about ninety or so undergraduate students hired each summer. Goatlike, nearly a hundred of the future's brightest minds bent down and moved through the plots, removing, one by one, any plants that had not been planted, any that did not belong.

As this experiment was set up and Tilman began to consider its results, it became clear that a great many things could be predicted about a patch of field as a function of the number of kinds of plants it contained. In an average year, plots with more kinds of plants yielded more biomass, more of the stuff of life. They contained more kinds of insects, both herbivores and species bent on eating those herbivores. They were also less susceptible to the invasion of pests and parasites. Along with tens of students and collaborators, Tilman spent decades writing paper after paper, first from the 1982 experiment and then from the bigger experiment, nearly all of which might have been titled "The Effects of Plant Diversity on . . . ," where what differed was the noun at the end of the sentence. Meanwhile, he needed to wait to test whether in these big plots too, as in the earlier, smaller plots he had laid out, more diverse fields would also be more stable. Once again, he had to wait because he needed years of data to consider variation across both good years and bad.

There are two reasons that more diverse plots of grasses, forbs, trees, or even algae might be more stable. The first is what has come to be called the portfolio effect (or the insurance effect). "Portfolio effect" is a term first used by stock investors. Investors who put their money in stocks from different industries or companies with different niches lessen their risk. Different industries respond to the same economic shocks differently. Some rise, some fall. As a result, investing in a diverse portfolio of stocks buffers risk and, in the long run, yields a higher average gain in value. The portfolio effect in ecology is similar. The more species there are in a given patch, the greater the odds that, for whatever new condition might appear in the future, at least one will be able to do well. So too the greater the odds are that one particular species might be an especially good performer (ecologists call this phenomenon the sampling effect). The portfolio effect is predicted to be especially pronounced if the diverse species tend to have very different niches. Imagine two scenarios, one with two plant species that differ subtly in their drought tolerance and susceptibility to flooding and a second in which one plant species is very drought tolerant and the other very flood tolerant. It is in the second scenario that the portfolio effect is greatest.

The second explanation is related to competition. It assumes that the species with the ability to live in the new environment, whatever it is, not only survives but also takes over the resources that other species were using. We can imagine this occurring over two years. In the first year of a new condition, the species with the right traits for that condition are more likely to survive. In the second year, they survive, reproduce, and take over some of the land on which the others once grew. In the context of stocks, having species that have different niches and yet, to some extent, compete is akin to investing in the stocks of both a company that produces solar panels and one that mines coal. They respond differently to economic and social changes, yet if one fails the other has an opportunity. Because the effects of competition require more time to become apparent, they are more likely to affect the resilience of ecosystems than their resistance.

Tilman finally compared the effects of the diversity of his big plots on their stability in 2005, ten years after setting up the experiment. He found that, just as in the previous experiment, plots with more species varied less, year to year, in response to the vagaries of climate and other factors.[5] In those plots, even if one species was susceptible to a particular hardship, others were not. The effects of their loss were buffered by the broader portfolio of species. In dry years, for instance, the drought-intolerant species withered, but the drought-tolerant species did not. When parasites swept through, the parasite-susceptible species died; others did not. But in plots with just a single species or only a few species, there was no such buffering. Some of the low-diversity plots did better in response to a particular problem (if, for example, they were full of drought-tolerant species during the drought), but on average they did much worse. Finally, if hard conditions lasted long enough, competition mattered. The drought-tolerant species, for instance, took over turf from those that were less tolerant.

ECOLOGISTS DO EXPERIMENTS in order to disentangle cause and effect in nature. They manipulate a few factors in old fields, ponds, or, when even ponds are too big, kiddie pools filled with algae and, say, tadpoles while keeping everything else the same. Implicitly, each ecological experiment is a microcosm of the world as a whole, the macrocosm. Ecologists look down on their microcosms, their miniature worlds. They tweak their conditions, rearranging the pieces of the living world. They then step back to see what results.

If everything goes well, they use the insights they have gained from those results to see the real world, the whole world, in a new light. For as much, then, as Tilman seems to have enjoyed working among his square plots of grasses and forbs, understanding their details and dynamics, he also imagined that he was making sense of the living world more generally. He was studying whether small plots of grass that are more diverse are also more stable. But he was using the results of those studies to predict whether whole habitats or even countries that are more diverse are also more stable. Ultimately, it is the former question that is answerable, even though the latter motivates the inquiry. But often the former becomes so encompassing, so like a world in and of itself, that the latter never ends up being discussed, much less well understood. Yet it is worth noticing that in some ways, each of Tilman's plots with more or fewer species was a microcosm of something bigger, a bigger grassland or a forest or even a country.

Tilman's work, if scaled up, implicitly leads to the prediction that diverse forests should be less susceptible to catastrophic pest outbreaks. The forests should be more stable and have a higher mean productivity. This appears to be the case, at least for temperate forests in Japan.[6] Countries with more-diverse forests should also receive more-stable services from those forests, whether with regard to water purification, pollination, the sequestration of carbon from the atmosphere, and much else. Countries with diverse grasslands should be less susceptible to sudden changes in the abilities of those grasslands to sequester carbon from the air (and help mitigate climate change) and, as a result, on average should sequester more carbon.[7] Surprisingly, as of yet, few of these predictions have been tested, and where they have been tested it has typically been at small scales, within patches of a particular habitat type rather than among states, territories, or countries.

There is also another prediction, the one perhaps most important to our own immediate future. Tilman's work also leads to the prediction that countries with more kinds of crops should be less susceptible to nationwide failures in crop productivity and all the societal consequences that such failures entail, which is to say, they should be resistant. This resistance, perhaps combined with resilience, should also tend to lead to a more stable food supply.

One could reimagine global agriculture in light of the insights from Tilman's plots, but doing so is not something that should be embarked on lightly. It would help if someone had studied the effects of crop diversity on the agricultural yield of whole countries. But as of 2019 no one had. One group of people who might have conducted such studies are climate economists. Climate economists have gathered enormous databases on the consequences of changing climate on societies. But the studies in such databases, such as those developed by the climate economist Solomon Hsiang, whose research was described in Chapter 5, tend to be focused narrowly either on particular components of modern societies, such as individual cities, towns, or even just buildings, or on ancient societies. Studies of ancient societies could, potentially, allow for tests of links between crop diversity and stability or crop diversity and collapse. But data on crop diversity are seldom available. Even when they are available, they are subject to debate. (I once sat through a morning of talks about the question, "How dependent were the Maya, *really*, on corn?") What is more, for those societies on which Hsiang and other climate economists have focused, the effects of climate changes seem to be independent of any details of those societies. Repeatedly, when climate changed, human societies suffered the consequences of that change. As Hsiang put it when I talked to him over the phone, "We see it again and again, a society is on top of the world and then climate changes, agriculture collapses, and the society collapses. Angkor Wat in Cambodia. The Mayans in Central America."

The results of studies of ancient societies by climate economists are grim. So too are the studies of modern societies at the edge of the niche of high-density human populations, populations dependent on agriculture. Superficially, such studies offer little hope for the value of crop diversity. But it may be that the effects of ancient crop diversity are difficult to see after so much time has passed. Perhaps diversity did benefit some ancient societies and allowed them to do better than they might have otherwise done, but such relative successes are lost to the teeth of time.

Inasmuch as Tilman's results from old fields are so unambiguous, we could always ignore the past and just implement his insights in the real world today. We could encourage regions, states, and countries to develop systems that favor the planting of more-diverse crops so as to ensure crop resilience and, in doing so, reduce the risk of the hunger, violence, and instability that can follow agricultural collapse. But it is a big jump to go from small experimental plots in Minnesota to the world. What works in an old field might not work for a region, much less at bigger scales.

Fortunately, with a little help, Tilman found a way to test his ideas at a bigger scale. Delphine Renard, at the time a postdoctoral researcher working at the University of California, Santa Barbara, teamed up with Tilman to test the diversity-stability hypothesis at the scale of the globe. Renard would focus on crops. All crops. Everywhere.

To get started, Renard sought out data on the crop species planted in each country, their relative abundance, and, importantly, their yield. The yield of a particular crop is related to the metrics Tilman used in considering old-field plants, their biomass, but it is slightly distinct in that it focuses only on the parts of crops—be they seed heads, fruits, or, more rarely, stems—we humans use. The national yield of any particular country, as Renard calculated it, is the total harvest of all crops (in kilograms) in a country divided by the area of cropland of the country (in hectares). Renard then converted those data on yield into a more intuitive metric, calories. Renard created a map of the total number of calories grown by farmers by country, by year, something like a nutrition label for the entire world.

In addition to estimates of the yield and calorie production of each country in each year, Renard sought out data on some of those variables that Tilman had measured for so very many years in old fields. Rather than collecting her data in the field, though, Renard worked her way through international databases. Rather than weeding out plants, she weeded out data sets that were unnecessary or not quite right. In the end, she compiled data for fifty years (1961 to 2010) for 176 crop species and ninety-one countries. It wasn't easy, but it was easier than the fieldwork Tilman and his collaborators had done over decades at Cedar Creek, easy enough that someone could have done it long before. No one had.

Renard's predictions built on Tilman's predictions for the old fields. She predicted that countries with more kinds of crops would experience fewer crop losses in years in which conditions changed, which is to say, they would be more resistant to year-to-year variability in climate and whatever else. Renard predicted that as a result of those reduced losses, they would experience less year-to-year variability in yield and hence would be more stable. In addition to crop diversity, Renard also looked at the effects of a handful of other factors that might influence year-to-year differences in yield. One was how much fertilizer was used. Another

Figure 7.2. The number of species of crops in different countries averaged over the ten years 2009 to 2019. Higher values, shown in darker shades, are countries in which more kinds of crops are planted. During the period covered by this map and Renard's research, Peru, Portugal, Cameroon, and China, for example, had high crop diversity. The diversity of crops in the United States and Brazil (where corn and soy, respectively, reign supreme) was, conversely, very low. Figure created by Lauren Nichols based on data from Delphine Renard.

was the extent of irrigation. She thought that fertilizer use might buffer year-to-year variation in conditions and that irrigation might do the same.

In a way, the problems Tilman considered out in his old fields were toy problems. He was in control of the experiments, and the stakes were low. His work was most relevant to the audience of other ecologists. Renard's analysis was very different. Droughts have been threatening agriculture around the world. There are new worries about a global food crisis, and as I discussed in Chapter 5, violence appears to be associated with climate-induced crop losses, particularly in extreme climates. Whatever Renard's results were, they would be relevant to the future well-being and survival of billions of people.

When Renard looked at the results, they were surprisingly clear. But to explain them I need to introduce one more concept, that of "evenness." Imagine a pie, an actual pie. It might be a key lime pie, for example, with a perfect crust and sweet and slightly sour filling. Now cut it into ten pieces. The pieces represent, in this analogy, crop species. If all the pieces are the same size, they are even. Conversely, if they differ in size, they are uneven. The most uneven cutting of the pie is one in which one piece is enormous (nearly the whole pie) and all the rest are tiny slivers. Renard integrated this concept of evenness into her thinking when considering the diversity of crops in a country. She calculated a measure of crop diversity that considered both how many crops were being farmed and how even they were. Let's just call it the diversity and evenness index. Renard's prediction was that a country could best buffer drought and other problems when its crops contained many species and those species were relatively even in the proportion of the land that they occupied—in other words, when its diversity and evenness index was high.

First, and unsurprisingly, Renard's results showed that one way countries could buffer climate variability was through irrigation. Countries where irrigation was more common more readily buffered dry years. Irrigation will continue to be important, particularly to the extent that it can be done more intelligently, based on real-time data on crop conditions and weather. Data-driven watering is about the least sexy-sounding solution to any problem in the world, but that is probably also what our ancestors said about the first stone tool.

But irrigation was not the only important factor. The diversity and evenness of crops also mattered. Those countries that had the highest crop diversity and the highest crop evenness indexes were more likely to have more consistent yields, year to year; their yields were more resistant. For example, the countries with the most diverse and even crops experienced declines in yield of 25 percent or more only very rarely, roughly once in 123 years. Conversely, countries with lower crop diversity and evenness had less consistent yields; their yields were less resistant and, as a consequence, also less stable. Countries with low crop diversity tended to experience a national yield decline of more than 25 percent once in every eight years. Importantly, the more stable yields of countries with higher diversity and evenness of crops were not associated with decreases in average yields. Countries with high diversity and evenness could have high average yield *and* high year-to-year stability.

There are many things we still don't know about crop diversity and crop resistance and stability, things we don't know and yet about which we might still make predictions. We don't know whether some kinds of crop diversity are better than others. However, the research done on Tilman's plots and elsewhere tends to suggest that the more different crops are with regard to their tolerance of the most common disturbances—for example, drought—the more likely it is that the success of one will compensate for the struggles of another.[8]

We also don't know the relative importance of the diversity of species of crops versus the diversity of varieties of those species. Figuring that out is important because while many countries and regions grow a diversity of crop species,[9] the diversity of varieties of those species has, for the most part, decreased.[10] Insights from old fields suggest that crop varieties are likely to be more important where societies are particularly dependent on individual kinds of crops for their sustenance (for example, cassava in sub-Saharan Africa and tropical Asia), whereas a diversity of crop species is likely to be more important where a variety of crop types are used as staples or where crops are produced primarily for export.

It is also unclear whether diversity buffers many kinds of year-to-year variability in conditions or just some kinds. Are countries and regions with a diversity of crops buffered both from year-to-year variability in precipitation and temperature and the arrival of new pests and parasites (the end of escape)? Research in Tilman's plots suggests that the answer is yes.[11] Then there is something else.

While the diversity of crop species has been stable or has even increased in some regions over recent decades, the crop species and varieties being farmed in different countries are more similar to each other than they have ever been. Some countries grow diverse crops, but they are the same diverse crops that other countries grow.[12] Research in old fields, kiddie pools, and other microcosms of the big world suggests that in such a scenario, a really bad year in one country is likely to be a really bad year in many countries. Whether this is the case will be hard to ever study at the scale of the globe. One must wait for a year so bad that even the agriculture of countries that grow many species of crops suffers. It is not a thing anyone hopes for, but if it happens, Renard and Tilman will, no doubt, analyze the resulting patterns of loss and hardship, searching for truths on the basis of which we can better manage the living world.

What we do know is about the effects of having a diversity of crop species at the scale of countries. If we imagine the future will include good years and bad years, as well as rarer and yet inevitable years of truly terrible drought, pest, and plague, then we are better off if, in the giant experimental plots that are our countries, we plant a greater diversity of crops. From other studies and from the traditional knowledge of farmers, we know that this is also true at smaller scales. Farmers can benefit from planting a greater diversity of crops. A field planted with more diverse varieties of rice, for example, is more resistant to pests and more stable in its yields than one planted with just one.[13] Similarly, a field in which more kinds of crops are planted (in rotation) over time is also more resistant to drought and more stable, over time, than one in which fewer kinds of crops are rotated.[14] This greater diversity is often harder for farmers to manage. It sometimes comes with a cost in terms of planting and additional challenges in terms of harvesting. Yet the more variable climate becomes and the more frequent and surprising crop pests and parasites are, the more important the benefits of diversity will become and the less important the costs.

The crow buffers its risk by knowing how to find different kinds of foods in different settings and conditions. Our tendency has been the opposite, to grow and eat the relatively few kinds of crops that have grown well in the past. But the future will not be like the past. It will be warmer and, in many places, drier (though in others, much wetter). It will also be more variable. In

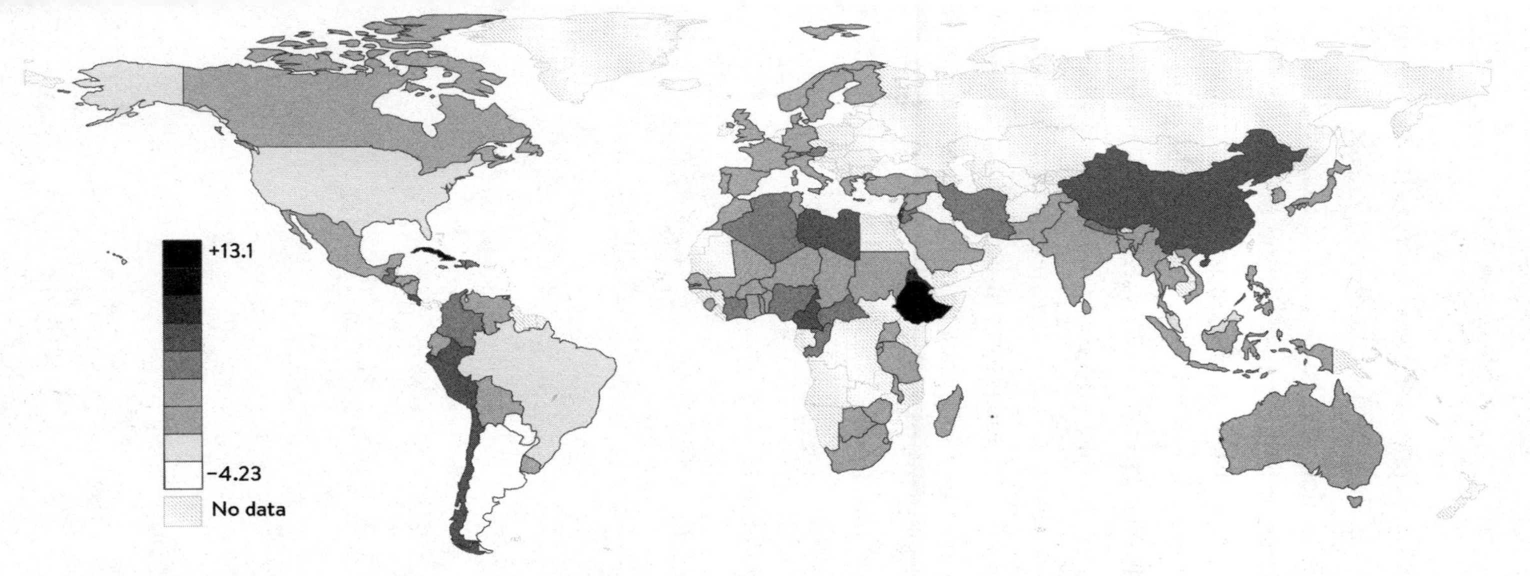

Figure 7.3. This map shows the change in the number of species of crops farmed in different countries over the past fifty years. Countries with darker shades now farm more species of crops than they used to; those in lighter shades farm fewer species. Measured at the level of species, crop diversity has declined in roughly half of countries in the last fifty years. In these countries, the potential for crop diversity to buffer climatic variability has been reduced. Conversely, some countries have increased their crop diversity. Those countries include Ethiopia, Canada, and China. Figure produced by Lauren Nichols based on data from Delphine Renard.

the future we will be better off if we grow diverse kinds of food so that we have something to rely on regardless of the conditions in a particular year. To achieve this goal, we need to have diverse crop species and varieties at hand to plant in the first place. The more variable and extreme future conditions become, the greater the diversity we will need. That diversity will need to be planted in farms, but it will also need to be stored in seed banks and other repositories. In addition, we also need to conserve the diversity of the wild relatives of crops, relatives that might help humans create more-diverse crops not today or tomorrow but, instead, hundreds, thousands, or even more years into the future. Nor are plants and their seeds all that we need to save if we are to buffer risk through diversity.[15] Not hardly.

CHAPTER 8

The Law of Dependence

IF WE MANAGE TO AVOID GLOBAL SOCIETAL COLLAPSE DURING THE next centuries, it will be because we figured out how to value the rest of life and the insights that arise from an understanding of that life. It will be because we realized we are dependent on the rest of life. There is no boundary between us and nature. We are as wild as we have ever been. Our very bodies—our skin, muscles, organs, and minds—are inseparable from nature. We are born of nature. This is a reality that has become very clear in light of Cesarean sections.

Cesarean sections, also known as C-sections, are relatively ancient in terms of human history. They are reported going back as far as 300 BCE, some 2,300 years ago. But like everything invented by humans, they are recent in the broader story of life. Between the origin of mammals, some 250 million years ago, and that first C-section, every one of our ancestors was born vaginally.

The earliest C-sections almost all involved removing a baby from a dead or dying mother. For example, the very pregnant mother-to-be of the second emperor of the Mauryan dynasty, in what is now India, was dying after having consumed poison. She is said to have undergone a C-section to save her baby. The baby survived and became emperor. The mother did not survive. Most of the many C-sections that were performed in the subsequent centuries were performed under similar, albeit not always so regal, circumstances. It was not until the early 1900s that C-sections became a common procedure that both mother and child were likely to survive.

Today, C-sections are still performed every day in order to save the life of a baby, a mother, or both baby and mother. They are also, however, performed as an elective procedure. It is the latter reality that has made C-sections very, very common. In the United States, 5 percent of babies were born via C-section in the 1970s. Today, one-third of babies in the United States are born via C-section.[1] The other two-thirds of babies are, obviously, born the old-fashioned way. The two groups then go on into their lives. As early as 1987, however, it had begun to become clear that babies born via C-section are different from those born vaginally, sometimes very different.[2] Some of those differences have to do with their body microbes. Body microbes are part of our body in the way that bees are part of a farm field; they are part of our bodily nature. Yet just as the bees in a farm field can go missing, so too can the microbes on our bodies, with potentially great consequences.

We have known about the importance of body microbes for more than a century. But much of what we know about body microbes we learned about first through the study of termites.

Termites are like humans in that they are social. They are like humans living in monarchies in that they have both a queen and a king. They are unlike such humans in that the queen is responsible for birthing, literally, her entire empire, egg by egg.

By the late 1800s it was already understood that some termite species depended on organisms in their guts in order to be able to digest wood (and especially the cellulose and hemicellulose of wood). Joseph Leidy, the progenitor of American paleontology as well as microbiology, broke open a termite of the species *Reticulitermes flavipes*, a species common in much of North America. He had been watching them "wandering along their passages beneath stones," and he "wondered as to what might be the exact nature of their food in these situations." So he dissected the intestines of a termite under a microscope. As he describes the moment,

> I observed brownish matter . . . contained within the small intestine. . . . [It] proved to be the semi-liquid food; but my astonishment was great to find it swarming with myriads of parasites, which indeed actually predominated over the real food in quantity. Repeated examination showed that all individuals harbored the same world of parasites wonderful in number, variety, and form.

Leidy called these life-forms "parasites," but he understood that they might be beneficial. He thought they were beautiful, and he and his wife drew them with what one could only call love. He also thought that many animals were probably inhabited, as are termites, by other species. He went so far as to say, "Some animals are so habitually and constantly infested with multitudes of various parasites that it would appear to be their normal condition."[3]

It is now clear that termites evolved from an ancient cockroach. They evolved, it is hypothesized, after a species of ancient cockroach began to live inside logs. The first termites lived in and consumed the logs. They did so in part through their reliance on single-celled organisms in their guts called protists. In termites, the protists carried out digestive processes that the insects could not carry out on their own. Then, while still inside the termites, the protists excreted compounds that the termites could more readily digest.

From the perspective of their microbes (including protists, but also other kinds of organisms, such as bacteria), termites offer housing and transport and a bit of food preparation to boot. They are an entomological mix between a taco truck and a bed-and-breakfast. The termites carry the microbes from place to place and supply them, constantly, with prechewed food. For the termites, the microbes are a necessity. Without their microbes, termites cannot eat wood. Without their microbes, termites are just roaches with big extended families. Without the right microbes, termites are dead, starved to death. As a result, termites must have a way of dependably acquiring the microbes they need.

Once it was clear that termites require specific microbes, it didn't take long for researchers to wonder how baby termites acquire those microbes. Answering this question is a bigger challenge than it might seem, since termites are born more than once. Termites hatch from eggs laid by the queen. They then grow by molting. Their old exoskeleton clouds and becomes transparent, and they break free from it, cracking their old self along its seams. They molt again and again, but rather than turning from, say, a caterpillar into a butterfly, they turn from a small termite into a larger one. Each molt is a rebirth. The problem with these molts, these rebirths, is that prior to each one, the termites' microbes are shed. After the molt they must be acquired anew.

The way termites deal with this repeated loss of the life-forms they need in order to carry out their daily functions, life-forms that together represent a kind of ecosystem in miniature, is to share them. Termites with microbes pass some of those microbes to those without by offering them hindgut fluids (a kind of special, microbe-enriched, fecal liquid) to ingest. When the colony is very small, this special form of provisioning is done exclusively by the king and the queen. This kind of proctodeal (from the Latin for "anus," *procto*, and "mouth," *odeal*) feeding seems gross. Yet the ritual sustains termite societies and replenishes their abilities to digest food that would otherwise be impossible to break down. This manner of feeding is a slightly complicated version of what evolutionary biologists call vertical inheritance. Termites inherit their genes vertically (their parents pass genes to them and they then pass those genes on to their descendants). The opposite of vertical inheritance is horizontal inheritance, in which animals acquire microbes (or genes, though that isn't our focus here) from the environment around them or from other individuals outside their family. It is thought that the ability and need to pass along microbes was part of what made termites social in the first place, unlike cockroaches. Termites needed to be around other termites, even late in their life, to reacquire their microbes. For this reason, they needed to be in a larger group, an extended family, a colony, a kingdom. Yet for as much as the dependence of termites and termite sociality on specific microbes has long been clear, the possibility that the same might be true of humans tended to be ignored.

THE TENDENCY TO ignore the human dependence on microbes was a mistake. Humans are no more or less dependent on microbes for survival than are termites. We depend on microbes for the development of our immune systems, for the digestion of our food, for certain vitamins, and as a layer of defense against parasites, among other things. More of the cells in our body are microbial cells than are human cells. The mystery is how we humans, or any other primates for that matter, acquire our microbes.

One hint as to the origin of our body microbes comes from the study of the microbiomes of wild primates, such as chimpanzees or baboons. For example, my collaborators and I studied the microbiomes of chimpanzees from thirty-two different wild populations across Africa. We were able to do so thanks to the work of a project called PanAf led by the Max Planck Institute for Evolutionary Anthropology in Leipzig, Germany. The PanAf project used camera traps to photograph chimpanzees and their behaviors. At the sites where chimpanzees were being observed by cameras, researchers collected feces after the chimpanzees left. Eventually, through a chain of intermediaries and next steps, my laboratory ended up with the DNA that had been separated out of those samples (which we then passed on to yet another laboratory). We discovered that the identity of the microbes found in the feces of the chimpanzees was a function of the group and lineage to which the chimpanzees belonged. In addition, the farther apart two groups were, the more different their microbes were. The microbes of the chimpanzees themselves were not solely influenced by the group to which they belonged and its geographic location. And yet the group to which chimpanzees belonged seemed to have the dominant effect on their microbes. Our results were very similar to those of my collaborator Beth Archie, a professor at Notre Dame University. Beth and her collaborators studied forty-eight baboons in and near Amboseli National Park in Kenya. They found that different groups of baboons had characteristically different microbes (similar to our result for chimpanzees) but also that even within groups, individuals that interacted more often shared more microbes.[4]

The similarity of chimpanzee or baboon microbes within groups has two very interesting facets. One is that it offers a potential advantage to individuals within the group. When individuals acquire the microbes of their social group, they are also more likely to acquire the microbes that work best given their social group's diet, environment, and even genes. In this way, as Beth Archie has argued, microbes of individuals can be, if not exactly tailor-made to local conditions, at least more tailored to those conditions than would be the microbes of an individual of a distant social group.[5]

But the other facet of the extent to which microbes of primates are similar within groups relates to how those microbes are acquired in the first place. This could result from food sharing, social interactions such as grooming, or even, just as with termites, eating each other's feces. But it could also occur earlier in life, during the messy process of birth. If this were the case, one might expect that babies should have microbes that more closely match their mother's than their father's or those of other community members.

Inasmuch as the lives of our ancestors were long similar to those of modern baboons and chimpanzees, it is likely that our ancestors acquired their microbes much as modern baboons and chimpanzees do. If that is the case, the study of human microbiomes might shed light not only on the human story, but also on the story of the acquisition of microbes in primates more generally (though it is also possible that the process differs from one primate species to the next). In humans, one can make and test some concrete predictions. If microbes are acquired from diverse sources in the social environment, the microbes of any baby (or adult) should be some complex function of the details of birth, early diet, social networks, and much more. They should be hard to predict. If, on the other hand, microbes are acquired during birth, then the microbes on vaginally born human babies should match not just those of their social group in general but also those of their mother in particular. Babies born via C-section might be expected to have microbes that come from other sources and that are also more variable.

ONE OF THE best-known recent studies of the differences in the microbes of human babies born vaginally compared to those born via C-section was led by Maria Gloria Dominguez-Bello. Dominguez-Bello grew up in Venezuela, before moving to Scotland to work on her PhD. She then returned to Venezuela, where she was hired at the Venezuelan Institute for Scientific Research. At the institute, Dominguez-Bello spent more than a decade studying the world of tiny species living in the guts of animals. When Dominguez-Bello began her career, most research on the microbes in vertebrate guts focused on domestic animals. Some of Dominguez-Bello's own research continued in this tradition. She studied sheep and cows, or rather, the life *in* sheep and cows. But Dominguez-Bello also began to study the guts of other species, species of the forests of her native Venezuela. Three-toed sloths. Turtle ants. Capybaras. Various smaller rodents. And then also the hoatzin, especially the hoatzin. Hers became a wonder-filled research program, research aimed at understanding the world of tiny, hidden forms and their abilities. No animal's gut was immune from her inspection, but for about a decade it was the hoatzin that especially attracted her attention.

Hoatzins are birds of the South American tropics with spiky "hair," blue eyeshadow, red eyes, yellow-tipped tails, and burgundy-fringed wings. They are high fashion and rock and roll all in one. But their striking visual appearance is not what is most unusual about them. What is most unusual about them is that, unlike most birds, they eat living leaves in great quantities. They ferment the leaves in a special gut that they and they alone have evolved. That gut is filled with microbes that the hoatzin uses both to break down the material in the leaves (much as the microbes in termites work) and to detoxify those materials at the same time. Dominguez-Bello started to study the hoatzin in the late 1980s, while still working on her PhD. She and her students and collaborators would go on to publish a dozen or more papers on the hoatzin and the unique ecology of its gut. She came to know the ecology of the gut of the hoatzin as well as anyone knew the ecology of the gut of any vertebrate.

Dominguez-Bello's career might have gone on like this, slowly lifting nature's curtain to reveal the marvels of the guts of animals on farms and in the rain forest. But then Hugo Chávez came to

Figure 8.1. A hoatzin, posing on a branch, its gut potentially full of hard-to-digest plant material and undoubtedly full of diverse species of bacteria able to metabolize those plants. Photo by Fabian Michelangeli.

power in Venezuela, and the hardships of his regime began to affect both daily life in Venezuela and scientific life. Dominguez-Bello left Venezuela for a new job at the University of Puerto Rico. With the new job, on an island far from home, confronted with new choices about what to study, Dominguez-Bello decided to consider human guts in more detail. She had studied human guts before. For example, she and her collaborators were the first to show that the stomach-dwelling bacteria *Helicobacter pylori* arrived in the Americas with the first Native Americans as they made their way from Asia. (Native Americans escaped some parasites and other bodily associates in their travels to the Americas, but not this one.) But now she would focus more exclusively on humans, while also continuing a few hoatzin projects on the side. It was in this transition that Dominguez-Bello came to wonder how baby humans acquire their necessary microbes.

Eventually, while based in Puerto Rico, Dominguez-Bello began to imagine studies aimed at understanding the details of how baby vertebrates acquire their necessary microbes. She envisioned two directions to study. One focused on her cherished hoatzins; it was to be the side project central to her heart. Working with Filipa Godoy-Vitorino, Dominguez-Bello compared the microbes found in baby hoatzins of various ages and in the crops of their mothers. She was able to show that some of the hoatzin's valuable microbes are shared, one generation to the next, much as they are shared in termites. When mother hoatzins feed their babies with regurgitated food from their crops, that food contains some of the mother's microbes. But those babies appear to also acquire additional microbes from their food as they grow, microbes from the surfaces of the leaves they ingest, microbes that make their guts ever richer with life over time.[6] The second direction would focus not on hoatzins but on humans.

Dominguez-Bello decided to study mothers and their newborn babies to compare the microbes of babies born vaginally with those of babies born via C-section, looking specifically at how likely it was that the babies' microbes matched their mothers. Here, she imagined the act of birth itself as being key to the conveyance of microbes. Perhaps, she thought, vaginally born babies tend to acquire their necessary microbes from their mothers' vaginas and skin or from feces expelled by their mothers during birth.

It was noted as early as 1885 that during a vaginal birth, babies ingest and inhale at least some microbes. It was also noted that they may also, and this was the word the scientists used, "onboard" them, via their anus.[7] These onboarded microbes can be microbes from the mother. In addition, newborn babies were noted to have the potential to onboard additional microbes (though in most cases, probably in lower doses) from their immediate surroundings, including other humans helping with the birth. Scientists knew these processes were occurring, they just didn't know how important they were. They didn't know if they were key to the process by which babies acquire the microbes they need in order to be healthy. Dominguez-Bello wanted to study microbes and birth in humans. This, anyway, was the long-term plan. But working with humans takes more planning than working with hoatzins, and this project was still in the planning phases. Then opportunity struck, in the guise of a transportation problem.

After completing her field research in Amazonas State, Venezuela, Dominguez-Bello had to wait for a helicopter out. Days and then weeks passed and the helicopter didn't come. She decided to take advantage of this "opportunity" of confinement to carry out her study on C-section and vaginally born babies. Because of another study she was leading, she already had the permits she needed. All she needed was permission from the local hospital in Puerto Ayacucho, which she quickly obtained. Documents in hand, and the helicopter nowhere to be seen, Dominguez-Bello set about recruiting mothers who would allow her and her collaborators to collect microbes not only from their own bodies, but also from those of their newborn babies. At the time the study was done, recruiting families for involvement in the work was hard and identifying the microbes present in individual samples was costly. Dominguez-Bello and her collaborators, as a result, decided to study relatively few babies: four born vaginally and six born via C-section. From those babies' mothers, the researchers took swabs of skin microbes, mouth microbes, and vaginal microbes. From the newborns they took samples of skin, mouth, nasal, and fecal microbes.[8]

What Dominguez-Bello and her colleagues found, once they had identified the microbes present on the swabs, was that the babies born vaginally tended to have more microbes typically associated with vaginal microbiomes. What was more, the microbes of an individual newborn tended to match those of its mother. Two mothers had vaginal microbiomes dominated by *Lactobacillus* bacteria. So too did their babies. One mother had a vaginal microbiome with more *Prevotella* bacteria (often also found in the gut). So too did her baby. The fourth mother had gut microbes that were from many different lineages. So too did her baby. This is like what one might see in termites, hoatzins, or elsewhere in nature.

But when Dominguez-Bello considered the C-section babies, she saw something different. The C-section babies at birth had microbes that were clearly distinct from those of their vaginally delivered counterparts. Their microbes tended to include species ordinarily found on skin, not inside the body. In addition, they did not reflect a signature of mother-to-child transmission. The microbes were not only not those of the baby's mother or family group. In some cases, they were not even microbes normally found on or in humans at all, except, subsequent research would reveal, other C-section babies.

Dominguez-Bello's initial research on birth considered a small number of mothers and babies. It was a kind of natural history, akin to Leidy's early work on termites, the sort of natural history driven by curiosity and wonder. It was this natural history that would help usher in the mainstream study of the acquisition of microbiomes by babies. It took a hoatzin biologist to remind medicine that we cannot consider ourselves separate from what Montaigne called "all other creatures."[9] We are connected to other creatures in two ways. First, we are more similar to other animals, such as termites and hoatzins, than we imagine. Second, we cannot be fully healthy without considering our dependence on other species, including microbes, among others.

Subsequent studies have refined the result of Dominguez-Bello's first paper on mothers and babies. We now know that the broadest of her results seem general. In general, vaginally born babies tend to acquire microbes from their mother, which helps those babies establish healthy gut microbiomes. Babies that are born via C-section acquire their gut microbiomes from elsewhere and can suffer from what has charmingly been called dysbiosis, a kind of collapse of the ecological community of the gut with sundry negative consequences. Subsequent research has altered our understanding of just how many of the microbes mothers pass to their babies are passed via the vagina itself and how many are passed via feces expelled during birth. A recent study led by Caroline Mitchell at Massachusetts General Hospital found little evidence of the establishment of vaginal microbes in vaginally born babies. It found, instead, strong evidence for the acquisition of fecal microbes from the mother by the baby during birth. Mitchell argues, convincingly, that a key component of this acquisition during birth may be not just that the baby acquires these microbes but that it acquires a big enough dose of them that they outcompete other species.[10] In addition, other studies have shown that other factors that might have an impact on the acquisition of microbes, or composition among microbes slightly later in life, also have an effect on baby microbiomes. This includes breastfeeding, which appears to help sustain maternally acquired microbes or, more generally, healthy human microbiomes. It also includes the use of antibiotics, whether by the mother before birth or the baby after birth, which tends to destabilize the microbiome and allow colonization of less than optimal microbes. These effects can persist into infancy and even adulthood.

We have also learned where many of the initial microbes of the C-section babies tend to come from. They come from the skin of the mother, nurses, and doctors but also from the air and surfaces of the hospital room in which the baby was born. They include unusual microbes, such as bacteria with the potential to cause disease and bacteria with genes associated with antibiotic resistance. What is more, scientists have also figured out, it seems, why some C-section babies come to have normal gut microbes and others don't. Some C-section babies, by chance, ingest fecal microbes from elsewhere in their environment. From dogs.[11] From the soil. From wherever those microbes might be found. In doing so, they acquire the microbes they need. But this chance pickup of necessary microbes has a statute of limitations, at least in humans. As babies get older, it is progressively more difficult for them to acquire new gut microbes, both because those microbes must compete with microbes that are already established and because the human stomach, while neutral at birth, becomes more acidic in the first year, as acidic as the stomach of a turkey vulture.[12] In addition, the later individuals acquire a healthy microbiome, the less likely they are to have the species they need during the critical early weeks, months, and years of development.

The dozens of studies that have by now followed up on the research by Dominguez-Bello differ in the details of their conclusions, but all agree on at least five points:

1. Vaginally born babies pick up many species of skin, vaginal, and fecal microbes from their mothers. Sometimes the microbes that colonize them nearly perfectly match those of their mothers. On other occasions, less perfectly. Caroline Mitchell and her colleagues, for example, found that in eight out of the nine families in which the analysis was possible, the strains of *Bacteroides* bacteria, the workhorse of the human gut microbial community, were exact matches to those of their mothers.

2. C-section babies tend to acquire their first gut, skin, and other microbiomes from the hospital room and the things in the hospital room.

3. For both C-section babies and vaginally born babies, other microbes continue to establish in the gut during the first year or two of life, a process that involves a succession of species and a gradual increase in diversity, the precise composition of which is guided by dietary shifts in the infant.

4. The microbes that babies pick up from hospital rooms are far less likely than the ones they would have picked

up from their mothers to be the microbes they need to thrive.

5. And, finally, babies who are born C-section but are exposed to vaginal or fecal microbes from their mothers can acquire healthy gut microbiomes, or at least microbiomes equivalent to those they would have acquired had they been born vaginally.

Just which problems do C-section babies potentially face by not being exposed to their mothers' microbes? They have the potential to suffer from basically any problem that is associated with not having the right microbes. These problems include a higher risk of diverse noncommunicable diseases, including allergies, asthma, celiac disease, obesity, type-1 diabetes, and hypertension.[13] It is also likely (though it has not yet been tested) that C-section babies are at increased risk for diverse sorts of infections, both because their own microbes are less able to defend them against parasites and because some of the species they acquire at birth are parasites.

These problems are diverse in part because the identity of microbes in a human gut and on a human body influence virtually every aspect of how the body works. Microbes aren't a key in a bodily lock. That is the wrong analogy. There isn't a single lock but, instead, many hundreds of locks, many hundreds or even thousands of ways and contexts in which microbes interact with our bodies. Individual microbe species might play more than one role and hence fit more than one lock. A single role might be filled by more than one microbe species. And which microbial key fits which lock depends on which other species are present on and in the body. All of this is to say that it is complicated. But just as importantly, we are naive. Most of the species living in and on human bodies have never been studied in any detail, despite having lived in, on, and alongside us for millions of years. We are still early in our understanding, and so it is not easy to identify, with any particular malady, what has gone wrong.

WE REQUIRE HUNDREDS and perhaps thousands of other species in and on our bodies in order to thrive and survive. In this, we are ordinary. All animal species are dependent on other species. This is the law of dependence. But animals also require a means of acquiring the species on which they depend, especially the microbes on which they depend. For some animal species, the microbes they encounter in their daily environments may be sufficient for their needs. For example, the ecologist Tobin Hammer has recently shown that the microbes in the guts of caterpillars tend to be those they ingest from the plants on which they feed. Similarly, baboons seem to be more able to acquire gut microbes from their friends after birth than are humans. But for many animal species, such environmental microbes are insufficient, so a kind of inheritance is necessary.

Termites accomplish something akin to vertical inheritance of their microbes, even in a context in which constant replenishment is required. Close relatives pass along the family microbes even when mom is far away. Nor are termites alone. Many animal species that rely on special microbes have evolved special ways of carrying them forward. Some beetle species have special microbe "pockets" on the outside of their bodies. Leaf-cutter ants carry their fungus in a little pouch beneath what would be their chin if they had a chin. Some insect species (many, really) go a step further in ensuring that the necessary microbes are passed on to their children. Carpenter ants, for instance, depend on bacteria passed by mothers to daughters, generation after generation, in order to produce some of the vitamins they require. At least one of those bacteria species is now housed, by the carpenter ants, in a special kind of cell that lines its gut. It is *inside* the ant's cells, integrated into its body. It is inherited by baby ants inside their eggs.[14] It is part of the ant's body, part of its egg, and yet it is still separate. Conditions too warm for the bacteria, but not the ant, kill the bacteria.[15] Then after a while, no longer whole, the ants slowly die too.

As we consider the future, one of our challenges is that we will need to find a way to continue to pass the species we need to the next generations. However, the need to pass our heritage on does not end with body microbes. The microbes passed from mother to child are just one small part of what is inherited. We are dependent on the inheritance of many species. Barry Lopez has written of the wolf that he is "tied by subtle threads to the woods he moves through."[16] We are tied by threads to much of the living world, through which our species has collectively moved. Let's imagine an extreme scenario to underscore the realities of the more ordinary scenarios. Let's imagine humans are able to colonize Mars. Among the scenarios for such colonization that have been discussed, there are two main possibilities. One is that we would colonize Mars with something like an enormous space station. A second is that we would colonize Mars and, using microbes of diverse sorts, reengineer its atmosphere to be more like that of Earth. Both scenarios are, for humanity, something akin to a rebirth, or at least a molt. By this I mean that they require us to take with us the species we need to survive. This is a much harder task than any that species on Earth must engage in. When a leaf-cutter ant queen flies to start a new colony, she carries with her the fungus that her progeny will grow on the leaves they gather. But she doesn't need to take with her the plants that make the leaves. We will need to take the plants and also much more.

We will need to carry with us the microbes able to break down human waste as well as the waste produced by whatever industries we would employ on the Red Planet. Right now, this is not done on the International Space Station. Instead, the astronauts pack their waste, fecal and otherwise, and bring it back to Earth, like fastidious campers. We will need to carry the species necessary to produce the food we eat. Individually, we consume hundreds or even thousands of species a year. Collectively, humanity consumes tens of thousands if not hundreds of thousands of species and far more varieties (the Svalbard Global Seed Vault, for instance, stores nearly a million varieties of crop seeds). In addition, such crops depend on the microbes they need, both on their leaves and on their roots. Many crop species, perhaps most, fail to thrive without their microbes. We could hope that crop parasites and pests don't arrive on Mars, but this is probably wishful thinking. If they do arrive, we need to be able to control them, and at least on Earth, the best bet for such control is often the enemies of those pests and parasites. The list goes on. But there is something else.

We can anticipate our needs today. But we cannot anticipate our future needs. As a result, our best approach is to keep around (and carry with us into the future) all the species we might need. Marie Kondo may advise us to keep our houses neat and devoid of many things. But she is only advising us about our own houses for our own lifetimes. We need to think about the world, and we need to think about the longer future. When we do, we need to keep the species that perform services for us today, but also those that might perform some service for us in the future. This is our ultimate challenge. The termite carries forward, from one generation to the next, its few cherished protists and bacteria. We must carry forward everything, the species we need today (which we only partially know well enough to list), the species we will need tomorrow, and the species we could need in the far future in any of the many worlds that might come to be.[17]

CHAPTER 9

Humpty-Dumpty and the Robotic Sex Bees

WHEN MY WIFE AND I WERE GRADUATE STUDENTS AT THE UNIVERSITY of Connecticut, we lived lives of relative frugality. What extra money we had was spent on plane tickets to Nicaragua and Bolivia, where we were conducting our respective research projects. As a result, when our vacuum cleaner broke, I took it on myself to fix it. Superficially, this was the cheaper solution. I took the vacuum cleaner apart without any trouble. I also identified the broken part. Then in trying to get the broken part off, I broke another part. Luckily, Willimantic, Connecticut, where we then lived, had a shop that sold vacuum cleaner parts and repaired vacuums. I bought the needed parts and went home, but even with all the parts in hand I could not put the vacuum cleaner back together again. I made one failed attempt, resulting in a vacuum cleaner that would suck air but sounded like a garbage disposal. I admitted failure and took the vacuum cleaner to the repair shop, disassembled, in a bucket. The owner looked in the bucket and said, without much fanfare, "Whoever tried to put this back together again was an idiot." In an attempt at saving face I blamed my neighbor, to which the repair shop owner said, "You need to tell your neighbor that it is easier to break something than it is to put it back together." He might have added, "especially if you aren't an expert." I bought a new vacuum cleaner.

That it is easier to break something than to put it back together or rebuild it from scratch is as true for ecosystems as it is for vacuum cleaners. This is a very simple sentiment, a sentiment that scarcely seems to rise to the level of a rule, much less a law. It is squishier than the species-area law, for example, and it isn't as direct a function of our senses as is Erwin's law. Nor does it have the same universality as the law of dependence. Yet it has enormous consequences. Consider tap water.

For the first three hundred million years after vertebrates dragged their big bellies up onto shore, they drank the water in rivers, ponds, lakes, and springs. Most of the time that water was safe. There were unusual exceptions, however. For example, water downstream of beaver dams often contains the parasite giardia. This parasite is unwittingly "contributed" to the water by beavers, in whom it often dwells, which is to say that beavers pollute the water systems they manage.[1] But as long as you didn't drink downstream of beaver settlements, for the most part parasites in water were rare, as were many other health problems. Then, just a moment ago in the big sweep of time, as humans settled in large communities in Mesopotamia and elsewhere, they began to pollute their own water systems, whether with their own feces or, once animals were domesticated, those of cows, goats, or sheep.

In those early settlements, humans "broke" the water systems on which they had so long depended. Until the cultural transitions that led to large urban centers, such as in Mesopotamia, parasites had been cleaned from water through competition with other organisms in the water and via predation by larger organisms. Most parasites were washed downstream, where they were diluted, sun-killed, outcompeted, or eaten. These processes occurred in lakes and rivers but also underground as water seeped through the soil and then into deep aquifers (it is into such aquifers that wells have long been dug). But eventually, as human populations grew, the water on which they depended came to contain more parasites than could be processed by nature. The water became polluted with parasites, which were then ingested each time someone took a sip. The natural water system had broken.

Initially, human societies responded to this breakage in one of two ways. Some societies figured out, long before knowing about the existence of microbes, that fecal contamination and illness were linked and sought ways to prevent contamination. In many places, this took the form of piping water into cities from more remote locales. But it could also include more sophisticated approaches to disposing of feces. In ancient Mesopotamia, for example, at least some toilets existed. Demons were thought to dwell inside those toilets, perhaps prefiguring an understanding of the microbial demons that fecal-oral parasites can be (however, there is also some indication that some people preferred to defecate in the open).[2] More broadly, however, approaches that successfully controlled fecal-oral parasites, whatever they might have been, would prove to be the exception. People suffered and were never quite sure why, a reality that continued, to varying extents in different regions and cultures, for thousands of years, from about 4000 BCE to the late 1800s, when the existence of a link between contaminated water and disease was discovered in London in the midst of what we now know to have been a cholera outbreak. Even then, the discovery was initially doubted (and fecal-oral parasites still remain a problem for much of the world's population), and it would take decades before the actual organism responsible for that contamination, *Vibrio cholerae*, was observed, named, and studied.

Once it became clear that fecal contamination could cause disease, solutions began to be implemented to disconnect urban fecal flows from drinking water. The waste of London, for example, was diverted away from the water that Londoners drank. If ever you feel smug about the cleverness of humanity, remember this story and its takeaway—namely, that it was not until about nine thousand years after the earliest cities began that humans figured out that feces in drinking water could make them ill.

In a few regions, the natural ecosystems around cities were conserved in such a way that the ecological processes carried out in forests, lakes, and underground aquifers could continue to be relied upon to help keep the parasites in the water in check. Communities conserved the natural ecosystems present in what ecologists call the watershed, the area of land through which water flows en route to some final destination. In natural watersheds, water flows down tree trunks, among leaves, into soil, between rocks, along rivers, and eventually into lakes and aquifers. In some places, the conservation of watersheds was haphazard or even inadvertent, the result of the idiosyncrasies of how cities grew. In other places, it was the result of the distance between cities and the places from which water was piped. In essence, water was kept safe by bringing it from very far away. In still other places, success came from investing heavily in conservation programs that ensured the protection of the forests around the city. This was the case with New York City, for instance.[3] In all of these scenarios, people continued to benefit from the parasite-controlling services of wild nature, often without knowing they were doing so.

In a few lucky regions, the services of nature are still intact enough to be sufficient or nearly sufficient to keep drinking water free of parasites. The far more common story, however, is one in which the water systems on which cities were dependent were not sufficiently conserved, or in which the scale of contamination and the disruption of the natural water systems proved to be too great for the amount of forest, river, and lake that was conserved. The great acceleration of human population growth and urbanization "broke" many rivers, ponds, and aquifers from the perspective of their ability to keep parasites in check. Independently, the people in control of different urban water systems decided that water would need to be treated, at large scales, to provide parasite-free drinking water to the urban masses.

Water-treatment facilities began to be developed in the early 1900s, and they employed a variety of technologies that mimicked processes that occurred in natural bodies of water. But they did so relatively crudely. They replaced the slow process of movement through sand and rock with filters, and the competition and predation of rivers, lakes, and aquifers with biocides, such as chlorine. By the time the water reached houses, the parasites would be gone and much of the chlorine would have evaporated. This approach has saved many millions of lives and remains the only realistic approach for most of the world. Many of our water systems, especially our urban water systems, are now simply too polluted to be relied on for untreated drinking water. In such contexts, there is little choice but to treat the water to try to make it safe again.

RECENTLY, MY COLLABORATOR Noah Fierer led a large group of other researchers, myself included, in a project to compare the microbes associated with tap water sourced from natural, untreated aquifers (such as that from household wells) to the microbes associated with water sourced from water-treatment facilities. Together, we focused on a group of organisms called nontuberculous mycobacteria. These bacteria, as their name suggests, are kin to the bacteria that cause tuberculosis. They are also kin to the bacteria that cause leprosy. They are not nearly as dangerous as either of these parasites, and yet nor are they innocuous. The number of cases of lung problems and even deaths associated with nontuberculous mycobacteria in the United States and a handful of other countries is on the rise. Together, our research group wanted to understand if these bacteria tend to be associated with either water from treatment plants or water that came from wells and other untreated sources.

Our team studied the microbes in tap water by focusing on a habitat where those microbes often accumulate, showerheads. What we found in studying the life in showerheads was that the nontuberculous mycobacteria, which are not very common in natural streams or lakes, even in streams and lakes contaminated with human waste, were far more common in water coming from water-treatment plants, especially water containing residual chlorine (or chloramine) meant to prevent parasites from living in the water during its trip from the water-treatment plant to someone's faucet. Generally speaking, the more chlorine present in the water, the more mycobacteria. Let me say this again for clarity: these parasites were more common in the water that was being treated to rid it of parasites.[4]

When we chlorinate water, or use other similar biocides, we create an environment toxic to many microbes (including many fecal-oral parasites). This has saved many millions of lives. However, this same intervention has also favored the persistence of another kind of parasite, nontuberculous mycobacteria. Nontuberculous mycobacteria turn out to be relatively resistant to chlorine.[5] As a result, chlorination creates conditions in which nontuberculous mycobacteria thrive.[6] As a species we disassembled a natural ecosystem and put it back together, more cleverly than I reassembled my vacuum cleaner and yet, nonetheless, imperfectly. Researchers are now working on ever-cleverer devices to be used to treat water, including ways to rid water systems of nontuberculous mycobacteria. Meanwhile, cities that invested in the conservation of forests and water systems and their services, and as a result rely less on water filtration and chlorination (or entirely do without), are in the enviable situation of having little in the way of nontuberculous mycobacteria in their tap water and showerheads. They have, in other words, one fewer problem to fix.

For hundreds of millions of years, animals have relied on the services of nature to reduce the abundance of parasites in water supplies. Humans, in producing large quantities of bodily pollutants and spreading them widely, overwhelmed the ability of aquatic ecosystems to provide these services. We then invented water-treatment plants to take the place of the natural services of aquatic ecosystems. But in doing so, we created a system that works and yet doesn't do all the things that its natural counterpart did, despite enormous investment. Something has been lost in the re-creation. In part the problem is one of scale (the great acceleration has led to a great acceleration in the amount of feces humans produce globally), but it is also a problem of our understanding. We don't yet know quite how forest ecosystems perform their services, such as those associated with keeping populations of parasites in check. Nor do we fully understand the circumstances in which they perform these services and when they don't. As a result, when we seek to engineer and re-create simpler versions of those ecosystems, we invariably make mistakes.

It is worth noting here I am not making an argument that it is necessarily cheaper to save nature than to rebuild nature. A large literature considers this sort of economic question, measuring things like (1) how expensive it is to conserve a watershed, (2) the net value of the services provided by that watershed, and (3) the negative long-term "externalities" associated with relying on a water-treatment facility rather than conserving the watershed. Externalities are those costs that capitalist economies tend to forget to figure into the calculations, such as pollution and carbon emissions. In some cases, many cases really, the ecosystem services provided by natural ecosystems are more economical than their replacements. In other cases, they are not. But this is not quite my point.

My point is, instead, that even in those cases in which the most economical (by any measure) solution is to replace a functioning natural ecosystem with technology, doing so tends to yield replicas of those natural systems that are missing parts and, more generally, act "like" nature systems but not as natural systems.

IN THE CASE of water systems, many cities had little choice but to embark on efforts to filter and chlorinate water. But if we look around, there are many new experiments in rebuilding ecosystems in which there are choices. The story of the pollination of crops in North America (and elsewhere) is one of those cases. Roughly four thousand native bee species can be found in North America. For millions of years, those bees pollinated tens of thousands of plant species. Then a series of unfortunate events occurred, unfortunate, anyway, from the perspective of native bees, native plants, and the future of agriculture. These unfortunate events occurred during attempts to rebuild farms and orchards in order to make them yield more food per acre.

Fields and orchards are to some extent replicas of grasslands and forests. The wild species in grasslands and forests have long provided food to humans. Farms and orchards provide food in higher quantities per acre and per year. These provisions are dependent on other species that live in the farms and orchards, or at least they were. Pests in fields and orchards were controlled by the pests' natural enemies. Wild pollinators helped pollinate the flowers of field crops and orchard trees. However, as the farming of fields and orchards became more intense, the pieces of the ecosystem began to be replaced.

The natural enemies of pests were, to varying extents, simultaneously killed and replaced by pesticides. In addition, heterogeneous farms, where many agricultural plant species were grown and where diverse native plants grew along the edges of rows, were replaced by monocultures, single species of plants grown at enormous scales. Monocultures of crops, along with pesticide use, led to changes in the service of pollination. Wild bee species need places to nest. Nesting habitats are rare in monocultures; each bee species needs a special soil type, soil structure, or plant material to build a nest. The soils and plant materials of monocultures are homogenized. Wild bee species also need sources of nectar and pollen throughout the season in which they are active. Monocultures tend to become food deserts for bees in the periods when the flowers on the crop are not present. In addition, wild bees suffer from the pesticides used to control pests. The pesticides don't know, for the most part, the difference between a weevil and a bee. As a result, there often weren't enough pollinators. Crops flowered but produced few fruits and seeds. The ecosystem had been reassembled in a way that left out a key piece.

The solution to this problem was to add another species to the ecosystem. In the 1600s, Europeans introduced a bee species called *Apis mellifera* to North America. Those bees, which we now typically simply call honey bees, are no more native to North America than are starlings, house sparrows, or kudzu. Yet as agriculture in North America intensified, honey bees became a key piece of glue necessary to hold together a broken agricultural system. Honey bees can be kept in high densities and then taken to fields when those fields have flowers that need to be pollinated. Beekeepers became a kind of sexual-service procurer for insect-pollinated crops. They fixed a part of the pollination system that had been broken, or at least partially fixed it. The challenge was scale.

In order to have enough honey bees to pollinate the broken agricultural system, the current solution is to farm honey bees across the country during the year (during which time they rely on wildflowers) and then, during the flowering seasons for different crops, drive the bees to where the crops are. At one particularly wild moment each year, for example, 2.5 million honey bee colonies are driven to California from all over the United States to pollinate almonds and other crops (but especially almonds). This system isn't great. It requires the bees to all be closer to each other, which predisposes them to sharing parasites. They pass a number of different bee viruses among themselves and also to native bees.[7] This happens in various contexts, but one of them is on flowers. To bees, flowers are like toilet seats. And while bees do wash their hands (or, rather, their feet), that is often not enough to prevent the spread of parasites. Viruses spread hive to hive. Protists spread. Even mites spread. But this is not all that spreads as part of the honey bee system; so too does a kind of genetic simplicity, simplicity and susceptibility.

Wild bees are genetically diverse. They are diverse overall in that they include many species. But they are also diverse in that each of those species also tends to include individuals with different versions of key genes. In addition, the wild bee species that are social tend to even have genetic diversity within their colonies. This diversity increases the odds that whatever parasite comes around will encounter at least some bees that are resistant, whether within a hive, within a species, or even just within an ecosystem.

The effect of diversity on the resistance of species to parasites was first studied in the context of crop plants. When farmers plant a greater diversity of crop varieties, the odds that all of them will succumb to a parasite decrease. The effect of diversity on resistance to parasites was next studied in David Tilman's plant biodiversity experiment, which I described in Chapter 7. In the fields of the biodiversity experiment, Charles Mitchell, now a professor at the University of North Carolina at Chapel Hill, showed that a plant parasite spread more slowly through more diverse plots than through less diverse ones.[8] Similar effects of diversity have since been shown within species. Patches of individual species of plants with more genetic diversity are less susceptible to disease. And we also know, now, thanks to the work of my colleague at North Carolina State University, David Tarpy, that honey bee hives that are more genetically diverse are less at risk from disease. Unfortunately, we also know that the honey bees within individual honey bee hives do not tend to be genetically diverse.[9]

In nature, honey bee queens mate with multiple males and, as a result, their progeny within an individual hive are genetically diverse. An individual queen might mate with eight or more males and, through her life, release their sperm into her oviduct to fertilize her eggs. Her offspring then have many different versions of genes related to resistance to parasites. However, the standard approach to honey bee management does not involve matings with multiple males. As a result, honey bees are genetically relatively homogenous such that a parasite that can infect and affect one bee in a hive can infect and affect most or even all of the bees. These homogenous hives are then brought together in enormous densities, where parasites abound. For part of the year, they are fed from just one food item, almond nectar. In bees as in humans, dietary homogeneity is often associated with ill health. Finally, the honey bees are often exposed to pesticides and fungicides in the fields they are asked to pollinate. The net result is that honey bee colonies are collapsing.

Native bees can play an important role taking honey bees' place, but in many agricultural regions they are now so decimated that their odds of replacing the services provided by honey bees are slim. Native bee populations were sufficiently suppressed by monoculture agriculture, pesticide applications, clearing of native grasslands, cutting of forests, competition with honey bees, and other assaults that they have struggled to bounce back. It isn't the native bee end-time, but it isn't a good time.[10]

What, then, is a farmer with a farm containing most but not all of nature's necessary parts to do? One solution that has emerged for some crops is to farm them indoors (or at least in greenhouses) and rely on boxes of bees, typically bumble bees, that are brought in and used precisely for the purpose of pollinating those crops. This is especially common with crops that are better pollinated by bees that vibrate at a specific bumble bee frequency, such as tomatoes. But it is also employed for other crops such as peppers and cucumbers. The approach is even more industrial than the approach used with honey bees, both in that the crops are indoors and in that the bees are used just for this service. They are not used for their honey. (Bumble bees do produce honey, but in small quantities in little tiny pots. It is too little to harvest commercially, though if you get a chance to stick your finger in one, take a taste; the honey can be lovely.) But bumble bees face some of the same problems as honey bees and have been more poorly studied than honey bees. They are harder to keep alive than are honey bees. And even when they are alive, they are trickier to manage than are honey bees. Their colonies are relatively short-lived. They don't survive the winter and rarely make it beyond a single season. As a result, farmers must purchase new bumble bees at least once and often more than once each year. Compare that to honey bees, which can be kept over the winter and for years, if cared for well. (Or compare it to wild bees that survive of their volition, so long as we don't destroy their habitats.) What is more, all the problems that face honey bees now will eventually confront bumble bees too. It is just a matter of time.

Most recently, a number of companies have begun to patent new robotic bees. These bees would fly from flower to flower, recognizing flowers via machine learning algorithms in their tiny robot brains, and pollinate them. Or at least in the future they might fly. The most advanced prototypes make do with tracks. They drive along the tracks to the flower, and then a little robotic arm reaches out. These driving prototypes are the size of dorm-room refrigerators and can presently pollinate a few flowers per hour while destroying a roughly equal number as collateral damage. I was going to say that they are analogous to sex robots for our crops, but they aren't *analogous to* sex robots. They *are* sex robots. These robots were invented to do something that nature already does, and the hope is that they would perform this service while roaming over hundreds of millions of acres of fields. Yet the approach is apparently appealing enough that Walmart has patented a version, though their patent is not for a functional prototype but, instead, for the idea for something that might someday work.

A field full of flowers and tiny robotic bees seems like an approach to the future as close to my bucket full of vacuum-cleaner parts as I can imagine. Bee biologists have been quick to point out, in their careful words, "This is f*$#ing crazy!" One group of bee biologists and pollination biologists felt sufficiently frustrated by the idea that they wrote a paper outlining the many ways this could go wrong.[11]

Many of the services wild species carry out and have carried out for millions of years face the possibility of replacement by some feat of human engineering. As we look to the future, many people are planning ways to replace the services of wild nature with technological alternatives. That is true, for example, of carbon sequestration. Hundreds of millions of years ago, plants evolved the ability to use the energy from the sun to bind together the carbon atoms found in carbon dioxide into sugars and thus create stored energy. All animal life would come to depend on this step. But then humans evolved and figured out how to burn ancient carbon in the form of coal and oil; in doing so, they released carbon dioxide into the atmosphere, which has triggered massive warming. New conversations are bubbling up in meetings around the world about technological approaches to gathering this carbon out of the atmosphere, quick fixes that replace the slow work done by plants. These may work marvelously. They may not. We would be wise to first learn as much as we can about the ways in which plants fix carbon, which communities of plants fix the most carbon, and how to save those communities. We would be wise to do so before, or at the very least simultaneous with, our attempts to engineer approaches to carbon fixation "faster than" and "better than" those carried out by nature.

The number of examples of cases in which we have attempted to fix what we have broken with technology goes on and on. Where we have killed off predators, we rely on people with guns to kill deer and control deer populations. Where we will have killed off the enemies of our pests, we must rely on the use of even more pesticides to keep those pests in control. Where we have cleared forests from the sides of rivers or straightened rivers, we must rely on levies and barriers to keep rivers at bay.

The more humans move indoors, the more remote the services of nature seem and become. The less obvious the services that nature is providing will be. The more normal "armies of plant sex robots" will seem. Similarly, returning to the example in Chapter 8, it is inevitable that we will try to find ways to simplify and replace our body microbes. We could figure out which genes we need from other species—the species in our guts, on our skin, and even in our lungs—and simply add those genes to the human genomes. Technologically, this is already possible (if cumbersome). But it will get ever easier. Genetically engineering humans is currently regarded as ethically problematic. But we are imagining the far future, and we have no control over the culture and ethics of our descendants. And so let's just imagine that genetically engineering people is among the range of things they consider. They might, for example, incorporate genes that allow human bodies to gather their own nitrogen from the air (as do some bacteria) or even that allow them to photosynthesize.

Yet digestion is trickier than nitrogen fixation or photosynthesis. The microbes in the gut talk to the immune system and the brain. They exchange the signals they have been exchanging for millions of years. We know that the specific details of these signals can affect how the immune system works (and when it fails) and also a person's personality. What we don't know is what the signals are. It has only been in the last few years that we learned that the signals existed in the first place. Maybe we'll figure this intestinal language out, decode each message, and figure out how to replace those messages with chemicals that send only the messages we want sent. Or maybe we'll figure out how to insert new genes into our cells to trick them into believing they have received a signal. The gut might send a signal, again and again, "I am pleased. I am full. I am pleased. I am full." But the hardest challenge remains our uniqueness. No two human genomes are the same; nor are any two human brains or immune systems. As a result, just what your body needs from microbes is not the same as what my body needs. Will we be able to tailor the genes that we edit to the person? Maybe someday.

Right now, replacing the roles of our bodily microbes with new cellular genes is an imaginary scenario, one predicated on what will be possible in the future. It is a scenario in which scientists become ever more clever and willing to manipulate nature, even human nature. But there is a second technological scenario. We could create seed banks for microbes and give newborn infants the microbes they need. We could also give adults the microbes they need when they have lost those they had. This is already happening in the form of fecal transplants, essentially a human form of the proctodeal feeding done by termites. Imagining a future in which newborns are colonized with microbes from microbial seed banks still requires that we know which microbes people need, given their own genes. But in theory this might one day be possible. My own prediction, if this is the route we go (and there are already such efforts), is that it will prove problematic for years, perhaps centuries, before it becomes reliable.

Ultimately, as we look to our immediate and distant futures, the easiest way forward is to conserve natural ecosystems and their services when we can. A second-best approach, and the one that we will often need to employ, is to find ways to mimic natural systems, to the extent we can, in ways that require as few additional interventions as possible. It is easier, to return to the example of gut microbiomes, to find ways to help mothers pass on their gut microbes to their children than it is to design the "perfect" mix of gut microbes from scratch for each child. The worst-case scenario is that of me and my bucket of vacuum cleaner parts; it is a scenario in which people around the world are left to solve the problems of the coming decades and centuries independently, without the insights that come from the experts, where the experts include engineers, ecologists, anthropologists, and people from other disciplines, but also nature itself. Of the ideas I've articulated in this book, the idea that we should save the services of nature, where we can, rather than trying to reinvent them is perhaps both the most obvious and the most contentious. It is obvious in that on some level it is intuitive that we should not break what is already working. It is contentious in that, increasingly, the future being imagined by scientists and engineers is one in which more and more of nature's services are replaced by technologies. Recently, a number of researchers have gone so far as to suggest that they don't need nature. They argue that, with genes in the lab, they can create whatever is needed. It is possible that they are right. I doubt it. I suspect my vacuum-cleaner repair person would doubt it too. And here is the thing: if they are wrong, and we have failed to save the ecosystems we needed, failed to keep them from breaking, well, then the consequences will be great. So I would suggest that the most sensible course of action is to proceed as though they are wrong and I'm right. Proceed as though the wild ecosystems on which we depend are not replaceable.[12]

CHAPTER 10

Living with Evolution

THE REASON WE TRY TO CONTROL NATURE IS THAT SOMETIMES DOING so proves extraordinarily beneficial, especially in the short term. When levees were built along the Mississippi, towns could be built along the Mississippi. Having those towns, some of which, such as Greenville, eventually became cities, so close to the river was a boon for transporting goods. It was, in the short term, a benefit. But it was a benefit with hidden costs, the costs associated with the floods that were to come. In the same way, when we seek to control the life around us by holding it back, we face similar realities. It can be beneficial to keep other life-forms at bay. We kill many species so as to make our own lives easier; we kill them to save ourselves. But such slaughter works best when our efforts are selective, when our attacks focus only on the species that really do us harm. When, instead, we try to kill everything, the consequences are predictable and inevitable. They flow into our lives like muddy river water.

During the great Mississippi River flood of 1927, which I introduced in the very beginning of the book, my grandfather told of spotting the place where the levee began to melt away just before the flood that poured through Greenville. He saw the levee bubbling. This story is both true and, almost certainly, untrue. It is true in that he may well have watched the levee bubbling and beginning to crumble. It is untrue in that once the river was high enough, the levee crumbled in many places, not just in the one spot where my grandfather happened to spy the water breaking through. At its height, the force of the river was simply much greater than the strength of the levee. Once the levee crumbled in one place, it crumbled in many. The river, in this story, is like life. The levee is our attempt to hold life back. The river cresting the levee, even pouring through it, is life reminding us at once of both its power and our weakness.

I think about that flood when I think about my grandfather. I also think about it when I consider the experiment I briefly mentioned in the Introduction, the experiment carried out several years ago by Michael Baym, Tami Lieberman, and Roy Kishony in Kishony's lab at Harvard University. The three worked together to design a giant Petri dish that they called the "megaplate" (in which "mega" was an acronym for "microbial evolution and growth arena" but also just meant "big"). The megaplate measured 60 centimeters by 120 centimeters by 11 millimeters. The megaplate experiment allows us to consider the nuances of one of the most rugged laws of biology, the law of evolution by natural selection, and to do so in real time. The law states, at its simplest, that the genes and traits of individuals that successfully beget more offspring tend to be favored relative to those of individuals that beget fewer. The law of evolution by natural selection is Darwin's law; it is a law that Darwin imagined played out relatively slowly but that we now know can work quickly. Its consequences can be seen in real time, whether in a city, on a human body, or in a megaplate.

The idea for the megaplate was inspired by a bit of movie marketing. In 2011, to promote the movie *Contagion*, Warner Brothers Canada created an advertising display in a shop window, a panel on which they grew bacteria and fungi in such a way as to spell out "CONTAGION."[1] The panel was in essence an enormous Petri dish. Kishony saw the advertisement and was inspired. This inspiration led, via conversations and a bit of brainstorming, to a similarly enormous Petri dish that was used in a class that Kishony was teaching and with which Lieberman and Baym, both graduate students at the time, were assisting. Like the panel, this Petri dish, the megaplate, would reveal a message. It was a message that would take longer to become legible, though it would eventually come to be just as clear.

The project took layers of teamwork. The whole team designed the experiment. Lieberman then carried it out for the first time with Kishony's class. Later, Baym fine-tuned the design and, in its final iteration, poured the agar, seeded the microbes, and watched what unfolded. The basic design of the megaplate was not dissimilar from that of the Warner Brothers sign, but with some key differences. For one, the agar that filled the megaplate was poured in two layers, a solid bottom layer that the bacteria could and would eat and a liquid top layer through which they could swim. A harmless strain of the gut bacteria species *Escherichia coli*, also known as *E. coli*, was then released from both sides of the Petri dish. The *E. coli* could both eat the nutrients in the agar and, because they can swim, move to places where nutrients were not yet depleted; they could dine and dash. Were there other species of bacteria in the Petri dish, the *E. coli* might not have

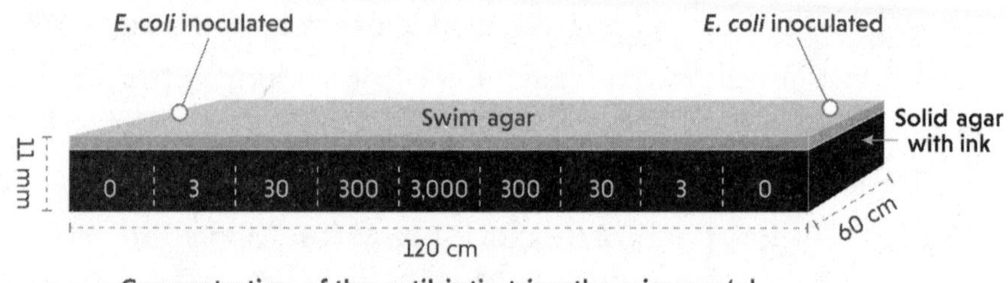

Figure 10.1. The megaplate designed by Michael Baym, Tami Lieberman, and Roy Kishony. This figure was produced by Neil McCoy based on an earlier version produced by Michael Baym and colleagues.

grown very well. They might have been outcompeted. *E. coli* is a useful laboratory organism, but it is not necessarily the toughest competitor when confronted with other denizens of the human gut. But this wasn't an experiment about competition with other species. It was about the evolution of resistance to antibiotics.

The bacteria released into the megaplate were not resistant to any antibiotic. They were susceptible, helpless even. But perhaps not for long. The team wanted to understand how fast these harmless, helpless *E. coli* might evolve resistance to a common antibiotic. How fast could resistant mutants appear and spread (even as nonmutants perished)?

In order to answer this question, the team decided to lace the megaplate with antibiotics. In the version of the experiment carried out by Baym after Kishony's class was done, the first antibiotic that was chosen was trimethoprim (TMP). Baym would later repeat the experiment with another antibiotic, ciprofloxacin (CPR), more commonly known as cipro. The antibiotics were not added uniformly to the megaplate. Instead, the plate was divided into columns. The columns were Lieberman's idea; she wanted to give the bacteria layers of barriers, each one in effect higher than the last. The outermost columns contained no antibiotics.

However, traveling inward from the edge, the megaplate contained columns with increasing concentrations of antibiotic until one arrived at the central column of the plate (equidistant to either end). The central column contained a concentration of antibiotic that should be sufficient to kill anything, a concentration that was either three thousand times (for trimethoprim) or twenty thousand times (for ciprofloxacin) higher than the concentration ordinarily necessary to kill *E. coli*. It is this columnar setup that reminds me of the Mississippi River near Greenville. The bands of antibiotics are levees. And the central band is, in this analogy, the town of Greenville, but in a broader context, humanity, humanity protected from the river of bacterial parasites by antibiotics.

To get to the central band, mutant bacteria would have to evolve resistance to the lowest concentrations of antibiotics. They

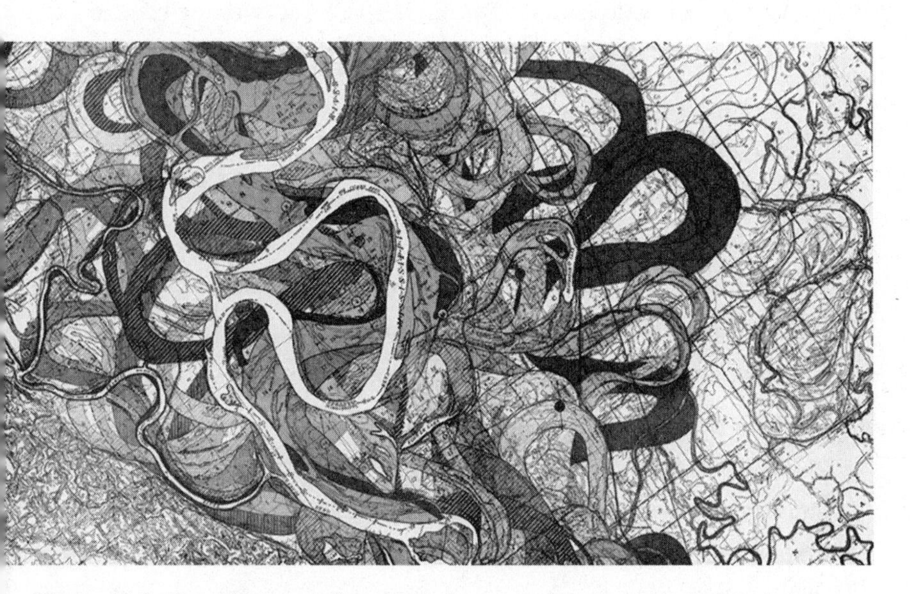

Figure 10.2. The Mississippi River and the meanders it made before being channeled and bounded by levees. It always had and still has a tendency toward motion and evolution over time. Map by Harold N. Fisk, US Army Corps of Engineers, 1944, published as part of the Geological Investigation of the Alluvial Valley of the Lower Mississippi River.

would then have to evolve additional mutations (in addition to those first mutations) for dealing with the next higher concentration of antibiotics. They'd have to do this, one layer of mutation on the next, until they had the set of genes that allowed them to make it to the middle of the plate.

The megaplate experiment has become a new classic of evolutionary biology in part because it was perfectly suited to illustrating the dynamics of evolution. As Jonathan Weiner wrote in *The Beak of the Finch*, his marvelous book on the study of evolution in the Galapagos Islands,

> To study the evolution of life through many generations you need an isolated population, one that is not going to run away, one that cannot easily mix and mate with others and, by mixing, mingle the changes induced in one place with the changes induced in others.[2]

Baym, Lieberman, and Kishony had imagined and then created just such a situation, and in a way particularly attuned Weiner's last concern, that mixing and mingling.

In hospitals and other settings where antibiotics are commonly administered, such as pig and chicken farms, one of the ways bacteria can evolve resistance to antibiotics is through sharing genes with each other via a kind of cellular swap meet that biologists call horizontal gene transfer. In horizontal gene transfer, bacteria mate and exchange plasmids—short bits of genetic material. Such matings can occur even between unrelated species, species as distant from each other as a goat is from a water lily. The results of such matings are hybrids with new genes that allow them to carry out tasks they couldn't otherwise carry out on their own. This sort of mating is happening around us all the time; it is happening right now in your body even as you read. But at the beginning of the megaplate experiment, this could not happen. None of the bacteria in the experiment had genes for resistance to trimethoprim or ciprofloxacin. Bacteria cannot share what they don't have.

The only way for the bacteria in the megaplate to become resistant was for them to undergo chance mutations in the letters of their genetic code, generation after generation, and for some of those chance mutations to yield versions of genes that allowed them to resist antibiotics. Any individuals with such genes would be far more likely to survive in the presence of antibiotics. That this might happen is a kind of wonderful lunacy on the basis of which evolution by natural selection works. Our own genome evolved (and evolves) in precisely this way. But our own genome has evolved very slowly.

With bacteria, Baym, Lieberman, and Kishony imagined they might see the sweeping dynamics of evolution unfold over shorter time scales. They had good reason to imagine this might occur. For one, the population sizes of the bacteria on the megaplate were immense, such that even though mutations are rare in *E. coli* (about one mutation per billion divisions), many such mutations might accumulate on the megaplate. In addition, the generation time of *E. coli* in the lab is about twenty minutes, allowing natural selection to act on those mutations again and again. That meant that in a little longer than a day, just thirty-one hours, Baym, who was the one checking the plate, could observe some seventy-two generations, the equivalent of studying a human population for two thousand years, back to the birth of Christ and then some. In ten days he could consider 7,200 generations, the equivalent of more than twenty thousand years in humans, back to the birth of agriculture and beyond. Yet though twenty thousand years may seem like a lot, the changes that have occurred in our own species in that time are modest, nothing we would notice at a dinner party. In this context, how long would it take resistance to evolve? When their megaplate experiments began, for all Baym, Lieberman, and Kishony knew, it might take a month or longer still, a year. Or even many years.

IT DID NOT take very long at all. The results were easy to see because Baym had dyed the solid agar of the megaplate black so that the white *E. coli* were visible as they divided and spread.

In the case of the trimethoprim, *E. coli* filled the first column of the megaplate, the column without antibiotic, with ease. They ate, excreted, and divided and then swam away in search of more food, ate and divided and swam away. The black ink disappeared beneath the white accumulation of their single-celled bodies. This was no surprise. During this time, many mutants are likely to have appeared, but none could survive the antibiotics in the second column of the megaplate. There was no visible evidence of natural selection or of the evolution it might have caused.

But when Baym returned a few days later, he saw something different. After about eighty-eight hours the first mutant with the ability to survive the lowest concentration of the antibiotic appeared. One bacterial cell had undergone a mutation that allowed it to survive in the low concentration of antibiotics present in the column. The descendants of that cell quickly began to flood the second column on one side of the megaplate with life. They turned sections of the second column of black agar white. Then, as Baym watched, other mutants appeared, independently, in the second column of the megaplate. They appeared and began to eat, divide, and spread. Sped up, the bacteria filling the second column and obscuring the black agar look like water in motion, roiling water, bubbling water, floodwaters. They have water's inevitability and also its power.

In his book *The Origin of Species*, Darwin wrote that "natural selection is daily and hourly scrutinizing, throughout the world, the slightest variations; rejecting those that are bad, preserving and adding up all that are good; silently and insensibly working, whenever and wherever opportunity offers, at the improvement of each organic being in relation to its organic and inorganic conditions of life."[3] Here Baym had seen that work, not over geological time but, instead, over days. The *slightest variations* were due to mutations, mutations of just a few trifling genetic letters. Those mutations were good, at least in the conditions created by the presence of the low concentration of the antibiotic. But as Baym was about to see, natural selection was not done with its silent and insensible working.

Over the next few days, mutations occurred in a smaller number of bacteria cells, mutations that conferred on them the ability to survive in the higher concentration of antibiotics. Natural selection favored those mutants. They filled the third columns of the megaplate. Then the process repeated with the fourth columns of the plate. New mutants arose, even more resistant mutants. Those mutants filled the fourth column. Finally, after ten days, a handful of mutants arose with the ability to survive in the middle of the megaplate, in the highest concentration of antibiotics. They had breached the final levee. After ten days, the middle column of the giant Petri dish flooded with resistant life.

Baym studied the results of the experiment, talked to Lieberman and Kishony, and then, because it's what scientists do, he repeated the process all over again. Again, it took ten days for the bacteria to reach the center. He repeated the experiment using the other antibiotic, ciprofloxacin. This time it took twelve days for the bacteria to reach and fill the center of the dish. He repeated the experiment again (and again), and over and over again it took twelve days. The results differed from one antibiotic to the other, but only slightly. And more importantly, in both cases the bacteria evolved resistance to very high concentrations of antibiotics very quickly. Other scientists have since repeated versions of the study with other antibiotics and other bacteria. The results are similar, differing only with regard to how long it takes the bacteria to get to the middle of the plate. The message written by a marketing team on the Warner Brothers sign that first inspired Kishony was "CONTAGION." The message written on the megaplate seemed to be more ominous; to Baym, Lieberman, and Kishony it read, "RESISTANCE."

In an evolutionary war with our microbial enemies, we are outmatched. This is true of the bacterial parasites and viral parasites on and in our bodies. It is also true of the species that try to eat our food before we do. Our enemies have an advantage because adaptive evolution is more rapid for organisms with big population sizes. The larger a population, the higher the probability that one individual will happen to have a mutation that is beneficial in novel conditions, such as in the presence of antibiotics, herbicides, or pesticides. The species with which we are competing also have an advantage because they tend to have short generation times. In each generation, natural selection has a chance to act. The more generations occur, the more readily it can favor some lineages, including new mutants, relative to others. The species with which we are competing also have an advantage because in the simplified ecosystems we create, they have little competition and few predators. They escape, released from their enemies, free to focus on devouring the food at hand. Finally, the species with which we are competing have an advantage thanks to our behavior. The more we try to kill them, the faster the rate at which resistant strains outcompete susceptible ones. Our best weapons disadvantage us.

So long as the human species is extant, the greatest opportunity for the evolution of new species will be in our farms, cities, homes, and bodies. These are and will continue to be the fastest growing habitats on Earth, and with their growth comes evolutionary opportunity for the origin of species. We are living with evolution.

The species that evolve alongside us in our daily habitats have the potential to benefit us, or at least, like crows, to live benignly alongside us, sharing our daily world. However, they needn't be so benign or charming. In all likelihood, they won't be. If we continue to try to control and kill the life around us, we will favor the origin of a very specific set of species, species resistant to our antivirals, vaccines, antibiotics, herbicides, pesticides, rodenticides, and fungicides. If we aren't careful, the species that evolve alongside us will all be dangerous, the garden of malevolent forms engendered by our attempts at control. Medusa turned all those who looked at her into stone; we turn those species we touch with our weapons into our near-immortal enemies.

The future need not unfold this way. The details of the evolution of the species that evolve in response to our assaults are often predictable. Where evolution can be predicted, we can use that predictability to our advantage. We don't have to wait for our bodies to evolve, one generation at a time, in response to the evolution of resistant parasites or wait for geneticists to breed or engineer new crops able to resist their pests. We can plan for the future through the application of what we know about evolutionary biology, or at least, we have the potential to do so.

But before we consider how we might keep resistance at bay, how we might, in essence, work with the flow and dynamics of the river of life rather than against them, let's consider in slightly more detail what happens if we don't. To do so, let's return to the megaplate experiment. It is a synecdoche, a part that represents the whole. Since Alexander Fleming first discovered that some fungi produce antibiotics that might be co-opted by humans, it has been clear that eventually the bacteria we try to kill with those antibiotics would evolve resistance to them. Fleming said as much during his Nobel Prize speech in 1945. In 1945, Fleming already knew that "it is not difficult to make microbes resistant to penicillin." What he feared was that antibiotics would become so readily available that they would be used ineffectively, in ways that favor resistance.[4] And that is what has happened. In the version of the megaplate experiment occurring on our bodies, in our homes, and in our hospitals, antibiotic-resistant bacteria are common and in many (but not all) regions are ever more common. Hundreds of antibiotic-resistant bacteria lineages have evolved in response to the massive scales at which we have used antibiotics and to the extraordinary quantities of food that our bodies represent to bacteria. Each lineage evolves in a slightly different way, contingent on its local conditions, its genetic background, and the antibiotic to which it has been exposed. Bacteria can evolve resistance by making a cell wall that the antibiotic can't see or bind to. They can make a cell wall that isn't permeable to the antibiotic, that won't let it in. The bacteria can supercharge a kind of internal pump that pushes the antibiotic out of the cell (like bailing water from a boat). They can evolve changes in the proteins in their cell walls to which antibiotics bind. The bacteria can even produce and wield a kind of biochemical knife to cut the antibiotic into bits. Or they can mix and match these defenses. Just as no two snowflakes are the same, neither are any two resistant bacteria strains.

Nor is this story of resistance just the story of bacteria. Protists, such as the species that causes malaria, also evolve resistance. The global story of the evolution of malaria parasites to resist the antimalarial drug chloroquine looks, on the world map, like the megaplate experiment writ large. Resistance evolved first in the mountains of Cambodia in 1957. Then the resistant form began to spread. It was able to outcompete other strains of the parasite anywhere that chloroquine was being used, which is to say almost everywhere. It spread to neighboring Thailand. Then it spread farther in Asia, then to East Africa, and then throughout Africa. Somehow, in the meantime, it had spread to the northern end of South America, from which it made its way to most of the rest of South America. It spread like the bacteria in the megaplate experiment. Meanwhile, as I write this, some strains of the virus that causes COVID-19 are beginning to evolve resistance to one or more of the vaccines.

Nor does the evolution of resistance end with microscopic species. The story of resistance in animals is much like that in bacteria or protists. Bedbugs have evolved resistance to half a dozen different pesticides, and it is estimated that no fewer than six hundred species of insects are resistant to at least one pesticide (and some are resistant to many). These include household pests but also crop pests. Crop pests evolve resistance to pesticides applied on fields and also to those produced by transgenic crops.

Evolution creates, and acts of creation are never complete because natural selection is busy bringing forth varieties, species, and forms. Through our actions, we shape those forms. We shape their identity, but also the details of their biology. As I've mentioned, the naturalist Buffon noted in 1778, "The entire face of the Earth today bears the imprint of human power."[5] That imprint favors the evolution of some species and disfavors others. We would be wise to try to favor the evolution of new worlds of lovely flowers, delicious fruits, or beneficial microbes. Yet this isn't our tendency. Our imprint is far more likely to favor new worlds of resistant forms.

BEGINNING IN 2016, I participated in a think tank on dangerous gardens of resistance.[6] The think tank was supported by the National Socio-Environmental Synthesis Center (SESYNC) and led by Peter Jørgensen, a scholar at the Stockholm Resilience Centre, and Scott Carroll, a researcher at the University of California, Davis. One of the necessary initial tasks of the think tank, called "Living with Resistance," was to understand whether the use of the biocides leading to resistance was increasing. You might imagine that it is someone's job to keep an eye on such things. It is not, at least not holistically. And so we tallied how many kinds of biocides were being used, how much was being used, and how widely they were being used. When we did, the picture was clear.

The use of biocides is increasing in tandem with the broader acceleration of human effects on the rest of life. This increase has many dimensions. For example, the number of doses of antibiotics being sold is increasing, as is the number of doses of antibiotics being sold per capita. Also, the number of liters of herbicides used is increasing, as is the number of liters of herbicide used per acre, as is the number of crops planted in transgenic herbicide-resistant crops on which herbicides are sprayed. The one biocide whose use has decreased is pesticide. This, however, is deceptive. The use of pesticide has decreased because reliance on transgenic crops that produce their own pesticides has gone up. The biocides used in chemotherapies to kill human cancers are also becoming more common. Cancer may seem very different from bacterial parasites or insect pests, and yet cancer can and does evolve resistance to chemotherapies, resulting in what are described as "non-responsive tumors," tumors that resist our attempts at control.[7] Nearly the entirety of the living world bears the print of human biocides. We have pressed our wide thumb ever more forcefully into the spinning clay of nature.

In most of these cases, resistance is becoming more common too. When we take antibiotics, our bodies become fleshy versions of the megaplate. We take antibiotics, and the bacteria evolve resistance and soon enough resume their growth, unbothered. When we dose our farm animals with antibiotics (often to spur their growth rather than to fix any medical problem), they also become like the megaplate. Bacteria evolve and grow in them, amid the swoosh and swirl of antibiotics, unbothered. Even our hospitals are like the megaplate. In hospitals, antibiotics are used on many patients and in many rooms. In addition, many people

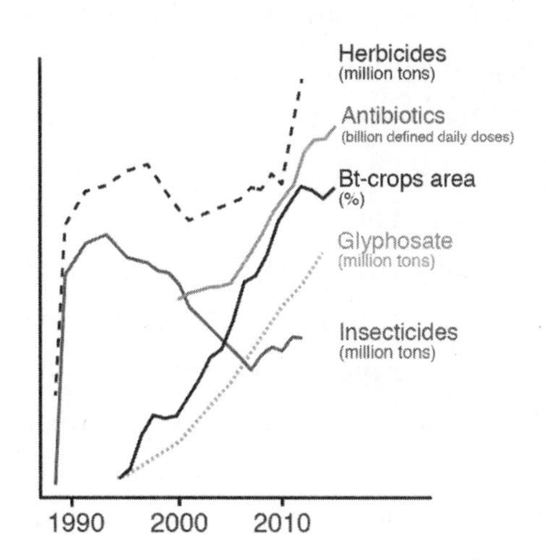

Figure 10.3. The change in the total use of herbicides, antibiotics, transgenic pesticide-producing crops (Bt-crops), one herbicide (glyphosate, sold as Roundup), and insecticides globally since 1990. Data are from Jørgensen, Peter Søgaard, Carl Folke, Patrik J. G. Henriksson, Karin Malmros, Max Troell, and Anna Zorzet, "Coevolutionary Governance of Antibiotic and Pesticide Resistance," *Trends in Ecology and Evolution* 35, no. 6 (2020): 484–494. Figure designed by Lauren Nichols.

in hospitals are immunocompromised so that their bodies are as defenseless as the agar in the megaplate. And the sagas of the evolution of cancers in our bodies unfold as though we were Petri dishes. Resistant cells, strains, and species grow, unbothered, throughout our societies' ecological systems. But "unbothered" is not the right word, because these cells, strains, and species actually do better in the presence of our biocides; their competition has been killed off. They grow as if favored, selected by us when we selected against the rest of life.

In general, the way we have prevented disaster, in each of these cases, is by finding ever-newer antibiotics, pesticides, herbicides, chemotherapies, and other biocides. As the evolutionary river rises, we make the levee higher. At first, we just searched for new biocides in nature; like gold miners, we prospected. We searched the biological world. This prospecting began long before Fleming or even the discovery of bacteria. For example, the medieval scholar Christina Lee and her colleagues have recently discovered an ancient Viking treatment for eye infections. Not only were Lee and her colleagues able to show that this treatment works to kill bacteria associated with eye infections, they found it also kills bacteria already resistant to some antibiotics.[8] (In other words, this ancient treatment is still useful.) A next phase of antibiotic discovery focused on invention, laboratory-based approaches through which scientists strategically create new compounds that might be useful. Now, desperate for antibiotics, scientists are studying how to mix approaches, the medical equivalent of the kitchen sink, an approach that relies on searching nature, searching traditional knowledge (such as that of the Vikings), and pure invention. Kishony, for instance, has helped pioneer a new approach, informed by his understanding of bacterial evolution, in which multiple antibiotics are used at once to treat infections. If done right, this approach can make it difficult for bacteria to evolve resistance to any one of the drugs, much less all of them.

The realities of resistance seem grim. But I'll offer a reason for hope. I have, by now, watched the video Baym filmed of *E. coli* evolving resistance hundreds of times. I show it in talks. It makes people quiet. It is what Kant called the horrifying sublime. But Baym thinks that the way people tend to view the video is wrong. He is far less terrified than you might be by what he has filmed. He is actually hopeful for the future of managing resistance, if we can take four steps. With each of these steps it is worth remembering that, just as with climate change, there is a time lag between what we do and how the world responds. We use biocides, and the consequences of that use turn up at some point in the future. But, unlike with climate change, that time lag is relatively short. It amounts to years rather than decades and in some cases is even shorter than years. As a result, it is possible to very rapidly make radical changes with regard to the ways we are managing the evolution of our enemies. In this context, these four steps become all the more important, because we can enact them now and feel the benefits soon. If we carry out these steps, we will dramatically increase our ability not to rid Earth of resistance (we can't) but, instead, to find ways of living with resistance, living with the flow and tendencies of life.

THE FIRST STEP to living with resistance is important, though little studied. It involves the idea of ecological interference. It is thought that resistant bacteria are less likely to be able to establish in conditions in which they are confronted with competition from other bacteria (many of which produce their own antibiotics) as well as with parasites and predators of bacteria. The more your hospital or skin is like a jungle, the less likely it is that any newly arriving bacteria strain will take hold.

The idea that parasites and pests might be less likely to thrive in the context of the diversity of their enemies is another law of diversity. It is a law tested in the old fields tended by Dave Tilman in Minnesota, the fields I described in Chapter 7. But there is a little more to this story in the specific context of resistance. Bacteria and other resistant organisms typically rely on specific genes that confer their resistance. Those genes are often big. It requires energy to copy them, again and again. The bacteria spend so much time copying them that they don't eat enough. In addition, the proteins and other products of those genes are often also costly in one way or another. As a result, resistant species are thought to be particularly susceptible to competitors and parasites. The first way to ensure that resistance is rare is through managing the ecological systems around us for diversity, when it is possible. This is something you can do at home: Use soap and water. Don't overuse antibiotics. Avoid hand sanitizers. Don't use pesticides when they aren't really necessary. All of these measures help preserve the beneficial species with which resistant species and strains compete.

The second important step to living with resistance is to manage our ecosystems so that they are more dominated by susceptible strains of the species with the potential to evolve resistance. This step is related to the first step. In the first step, the susceptible species are often competitors. We need to favor susceptible competitors. But susceptibility is important in more contexts than just competition.

A special version of managing for susceptibility occurs in the context of transgenic crops that produce their own pesticides. Such crops are safe to eat. But they are very susceptible to the evolution of resistance to the pesticide they produce. The susceptibility is problematic because the area planted in such crops is immense. As a result, when resistant pests evolve, they can eat from one field to another. They can devour whole countries. This has happened. It will continue to happen. But there is a solution, or at least a temporary fix, a way to put the problem on hold.

If plants that do not produce pesticides are planted near crops that do, pests will preferentially devour the defenseless, pesticide-free, crops. These pesticide-free crops are called refuge crops: they provide refuge to susceptible pests. In such situations, resistant pests might evolve, but any individual resistant pests will be most likely to mate with the more successful pesticide-susceptible individuals that are feeding on the refuge plants that do not produce pesticide. The genes for resistance stay rare in the pests, watered down by the genes of the more numerous susceptibles, especially if the genes for resistance come at some cost, as they often do. This approach might seem outlandish, but it works. In most countries that plant transgenic crops that produce their own pesticides, it is mandatory to also plant such susceptible refuge plants. Where it is mandatory and enforced, the evolution of resistance has been forestalled, and the value of the transgenic crops has been preserved. Where it is mandatory but not enforced, resistance has begun to evolve and "miracle" transgenic crops are being devoured, miracles no more. For example, in Brazil resistance is evolving even to the most well-defended transgenic crops. If this continues, Brazil will have to switch back to older agricultural systems (which require different seeds, different equipment, and much else) because new transgenic crops to replace those at risk are not likely to be released anytime soon. When we don't manage resistance well, the rate of our innovation does not match the rate of the evolution of resistance.

Approaches similar to the refuge-crop system have recently been advocated for the control of cancers in human bodies. For example, the evolutionary biologist Athena Aktipis, a member of our think tank, has set out a bold new approach to treating cancer in her book *The Cheating Cell*. Aktipis argues for treating cancers with chemotherapies only when they are actively growing.[9] If chemotherapies are used when tumors are not actively growing, the treatment kills off susceptible cells and leaves mostly resistant cells. Just like resistant bacteria, resistant cancer cells aren't good competitors, but with all the susceptible cells gone, they thrive. If the chemotherapy is used once, and is then used again before the tumor has begun to actively grow a second time, the last of the susceptible cells are killed, and all the cells that are left are resistant. When the tumor begins to grow the third time, the entire tumor is resistant. If, on the other hand, the tumor is treated only when it is growing, some susceptible cells survive because they divide and grow more quickly. As a result, the next time chemotherapy is applied, most of the tumor cells will be susceptible. This approach, part of what is called adaptive therapy, is part of new clinical trials being done by Bob Gatenby at the H. Lee Moffitt Cancer Center and Research Institute in Florida; so far, the trials have been very successful. Adaptive therapy isn't a magical solution to curing cancers; instead, it is a framework that can complement many existing approaches. It is an important start in terms of thinking about how to manage cancer resistance and how to work with natural selection rather than against it.

The management of transgenic crops and the treatment of cancer are different. Yet they share a fundamental component. In both cases, preventing the spread of resistant organisms depends on finding ways to favor susceptible organisms. Recently, the leader of our think tank, Peter Jørgensen, has argued that our enemies' susceptibility to our biocides is a kind of common good. It is, Peter suggests, a common good that is as important to humanity as, for example, clean drinking water. The more we manage pests, parasites, and even cancer cells so as to promote this susceptibility, the more control we will have against such species. Just how we manage for susceptibility differs from case to case, and yet keeping susceptible individuals around is something that benefits us all.[10]

The third step to living with resistance is trickier now, but it will be less difficult in the future. It relates to understanding the predictable features of the ways resistance evolves. In response to some biocides, organisms can evolve resistance in many different ways. Run the evolutionary tape back again and again, and it plays out a different way each time. But in other cases, the evolution of resistance is very predictable. Just what is predictable will differ from case to case. In some species, what is predictable is the speed with which resistance evolves. On the megaplate, resistance to one antibiotic evolved, again and again, in ten days. Resistance to the other took twelve. In other cases, more details are predictable. In some bacteria species, when resistance evolves to a particular antibiotic, it tends to involve the same mutations, in the same order, again and again, rote steps in an evolutionary dance. In such cases it is possible to anticipate these steps and get ahead of them. Here is a kind of precision prediction, predicting not just that resistance evolves but how, and then working to change things in that light. This will be possible in some species and kinds of resistance but not others. It is our job to figure out which is which.

The fourth step and last step to living with resistance involves a return to nature's solutions. It is an idea that Baym kept coming back to in our conversation, one that made him, he said, "hopeful." In my experience, biologists who study resistance don't use the word "hopeful" very often. Or they only use it ironically (or even sarcastically). And yet when I talked to Baym, he used the word and seemed to mean it. What made him hopeful was a group of viruses called bacteriophages.

In general, our biocides are hammers. Antibiotics kill bacteria, more or less without specificity. Pesticides kill insects. Herbicides kill plants. Fungicides kill fungi but also threaten many animals. Where biocides are specific, the specificity is coarse. For example, the most specific antibiotics tend to be better at killing either gram-negative bacteria species or gram-positive ones. That is, if there are a trillion species of bacteria, the antibiotics are specific enough to kill only half a trillion, not all of them. This is a dumb way to fight back against the species that threaten us. It is the equivalent of putting a moat around our civilization but not a bridge. As a result, the only species that make it into the castle are those that are tough enough to swim, climb the wall, and survive the boiling oil, and meanwhile, without a bridge, we have no way to escape once they arrive.

A more sensible approach would be to target specific foes strategically. Doing so requires us to know our enemies. Many common parasites are, as Baym put it, "in the stage of natural history." Some relatively common parasites have yet to be named. It wouldn't be hard to document our enemies more systematically; we just haven't, especially outside the most affluent countries. We need to know our enemies in general. But we also have to know the specific enemy that is affecting a particular patient; we need to be able to take a swab and identify the species on that swab, the strains of those species, and even the genes that strain possesses. A few years ago this would have been impossible. But now, not only is it possible, it has become much, much easier and cheaper. Soon it will be standard practice, at least in affluent hospitals in affluent countries, to know the entire genome of the parasite infecting someone. Once the parasite is known, it can be targeted not with a general antibiotic but, instead, with a bacteriophage specific to its genes and defenses. This approach won't be ready this year, but it will be ready, it increasingly seems, in the years to come. It is an approach that uses nature's diversity (of bacteriophages) to our advantage.

What all of these steps and their associated approaches share is that they require us to build on what we know about evolution's laws and general rules as well as on what we know about the detailed natural history and evolutionary tendencies of specific species. Our current practices of medicine and medical science are not particularly good at attending to either evolution or natural history. But we can improve them. The advantages to building medical and public health systems that rely on evolutionary insights and natural history are immense.

Perhaps we can change our ways. Michael Baym is hopeful. And the companies that are beginning to develop solutions based on the four insights I've outlined above are hopeful too. Maybe it is safe for you to be hopeful too, or at least not quite as pessimistic as the results of Baym's megaplate might otherwise make you feel, hopeful about our ability to make change. What will not change are the rules of evolution. Not in ten years or ten million, not until the end of life.[11]

CHAPTER 11

Not the End of Nature

IN 1989 BILL MCKIBBEN PUBLISHED HIS FAMOUS, FAR-SEEING, AND consequential book *The End of Nature*. The book was a rallying cry to fight on behalf of the future. It would help precipitate major momentum for conservation action, attempts to mitigate climate change, and more. And it would be followed by a series of similar books, most recently David Wallace-Wells's *The Uninhabitable Earth*. These books were important and useful, but they were also wrong.

What was wrong was not the idea that rates of change in the conditions for life on Earth were accelerating because of humans, that such change would lead to global human tragedies on a scale with little precedent, or that such change would cause increases in habitat loss threatening ecosystems, their wild species, and even the most basic services those ecosystems provide to humans. All of these things were and remain true. What is wrong is the idea that all of this has anything to do with the end of nature. Our end is far nearer than is nature's end. This is a reality that became clear to me in Okazaki, Japan.

I was invited to a conference on extinction. While finishing up my PhD dissertation in 2003, I'd begun to informally study the extinctions of insects. At the time, it was a lonely endeavor. I'd given a few talks about the list of insect species thought to have gone extinct in the last few hundred years. I'd spent tens of hours trying to document others of the insect species that were already extinct.[1] I wrote papers about the findings and made a website honoring those species. In collaboration with a graduate student in Singapore whom I had never met in person, Lian Pin Koh, I'd also started to study coextinction, the extinction of dependent species (such as mammoth lice) due to the loss of other species on which they depend (mammoths).[2] Somehow that collaboration led me to be invited to the meeting in Japan from Curtin University in Perth, Australia, where I was working at the time.

Luminaries of extinction research assembled at this meeting. Each was eager to present his or her own perspective on the big picture. Stuart Pimm, author of *The World According to Pimm*, talked about research attempts to estimate the global rate of extinction.[3] Robert Colwell talked about new ways of understanding where species were most diverse and why and how such knowledge might inform our understanding of extinction. Jeremy Jackson talked about the loss of big species in the sea and the ways each new human generation comes to accept some slightly smaller set of species as "big," some slightly lesser nature as nature. Russell Lande talked about the decline of small populations of rare species. If there was an overall feeling to the talks, it was that although precise estimates of extinction rates are difficult, the world was in trouble. Nature was in trouble. At the time this was not surprising. But hearing talk after talk about the struggles of wild species moved quickly from disheartening to demoralizing. Then Sean Nee stepped up to give his talk.

Sean Nee was at the time a professor at Oxford College. Though young, he had already gained a reputation as a clever iconoclast among evolutionary biologists. He saw and pointed out things other people missed. Sometimes he used math as the lens through which to see such things. In other cases, he just paid attention. This was one of those other cases.

Nee, as I recall, began his talk by showing an evolutionary tree of life, basically a family tree but for all the species on Earth. This tree bore little resemblance to the evolutionary trees in textbooks, most of which tend to focus on particular organisms of interest. You might find an evolutionary tree of humans, apes, and our extinct relatives, or an evolutionary tree of oak trees (a tree of trees). What most people, even most evolutionary biologists, only rarely see is the bigger evolutionary tree, the one that includes not just primates or mammals or even vertebrates but also fungi, pinworms, and all the ancient lineages of single-celled organisms. There is a reason for this.

Figure 11.1 shows a version of the big evolutionary tree of life. If the branches were all labeled, you would quickly notice that the names on the branches are mostly unfamiliar. Some of the big branches on the tree of life, for example, include the Micrarchaeota, the Wirthbacteria, the Firmicutes, the Chloroflexi, or the even more cryptic "RBX1," Lokiarchaeota and Thorarchaeota. If you were to look for the branch that includes humans, you might struggle to find it. This isn't a mistake but, rather, a reflection of our own place in life's bigger picture. Much like the tree Nee showed, this evolutionary tree makes clear that most of the branches on the evolutionary tree of life on Earth are devoted to different kinds of microorganisms.

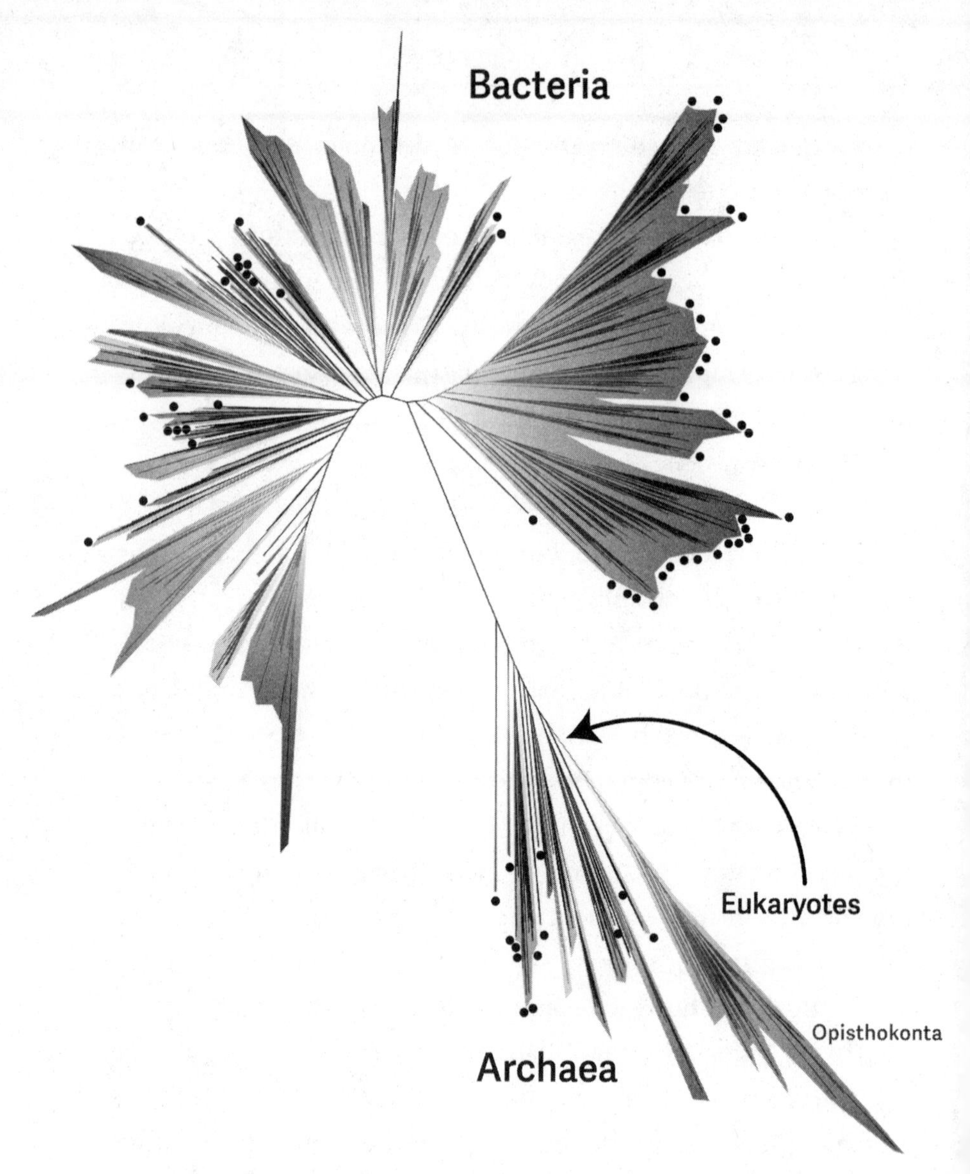

Figure 11.1. An evolutionary tree of life that includes all the major branches of life (but not all species!). On this tree, or rather, a sort of bush, each line represents a major lineage of life. All species with cells with nuclei are part of the Eukaryotes, represented as a single broom-like branch, indicated by an arrow, in the lower right-hand section of the tree. Eukaryotes include malaria parasites, algae, plants, and animals, among other life-forms. The Opisthokonta, one small part of the Eukaryote branch, is the branch that includes animals and fungi. Animals, if we zero in, are just one slender branch of the Opisthokonta. From this broad perspective, vertebrates do not get a special branch on the tree. The vertebrates are a small bud. The mammals are a cell in that bud. Humanity is, to continue the metaphor, something less than a cell.

We mammals can locate ourselves on the branch of the Eukaryotes on the bottom right of the tree. On the Eukaryote branch, we mammals are a tiny little bud off the twig called the Opisthokonta. As mammals, our distinctness is modest, and our branch is little more than a thin and unremarkable twig.

Biologically, the revelation in Nee's talk—that nearly all the ancient branches on the tree of life are lineages of microscopic, single-celled organisms—is not and was not news. This is a reality that scientists have known about for a long time. It is one facet of the Erwinian revolution (Chapter 1). It is a reality that began to be discovered when a microbiologist named Carl Woese developed a new way to study the life around us, an approach that allowed different life-forms to be compared in common terms, based on the letters of their genetic code. Until then, organisms tended to be compared based on what they looked like (their morphology) or what they could do (e.g., "grows in acidic conditions"). When Woese began to use his new approach, he was in for a surprise.

One of the samples that Woese was studying was of a bacteria species that looked like other bacteria and lived, like many bacteria, in cows. But as he considered this bacteria species genetically, he found it was different from the others. It was genetically more distinct from all the bacteria that had so far been studied as those bacteria collectively were from any other life-forms. After studying it, it became clear to Woese that this wasn't a bacteria species at all; instead, it was an entirely new kind of organism, an Archaean. The Archaeans appear in Figure 11.1 on the long branch that also includes us. Woese came to realize that the archaea, although they superficially resemble bacteria, are more similar to us than they are to bacteria. In addition, microbiologists, including Woese, would realize that many of the most ancient and unique lineages of life thrive under conditions so unusual to us that we have yet to figure out how to grow them in the laboratory. Each of the lineages marked in Figure 11.1 by a black dot is a lineage of life for which no species has ever been grown by humans. We know they exist because their DNA has been discovered and decoded. But we don't know what they need. Not only are these lineages of life unlikely to depend on us in any way, but we have yet to figure out what it is they need to exist. Some require extreme heat to thrive; others extreme acidity; others require specific chemicals released by volcanism. Many may well grow so slowly that what they most need is time; their metabolisms may be so slow that we can't detect their activity in the years provided by an ordinary human scientific career.

In making his argument, Nee was employing the insights of Woese and the microbiologists who built on his work. Nee's arguments would have been obvious to a meeting of microbiologists. Yet they were less obvious to conservation biologists. Nee brought microbiology to a conservation biology meeting, and in doing so, he called attention to one consequence of this evolutionary tree—namely, that if the diversity on Earth is measured in terms of ways of life, abilities to digest particular compounds, or even just unique genes, the better part of life is microbial.[4] Conversely, the mammals, birds, frogs, snakes, worms, clams, plants, fungi, and other multicellular species are, even when considered together, relatively insignificant.

As Nee pointed this out, the audience started to get a sense of where his talk might be headed. People began to fidget. The room grew a little quiet in anticipation. What Nee would go on to say was that all the worst things we can imagine doing to Earth—nuclear war, climate change, massive pollution, habitat loss, and all the rest—may affect multicellular species like us but are unlikely to lead to the extinction of most major lineages on the evolutionary tree. What is more, in the face of our worst assaults, many of the most unusual lineages would actually be more likely to thrive. On the first day of the conference, we had heard about the rarity of pandas, the endangerment of palms, and the critical population sizes below which species were unlikely to recover. It felt like the end of nature, but here was Sean Nee, arguing almost the opposite.

Some people were angry, but Nee was right in a way. Nature is not under threat. It is not ending anytime soon (which is to say, not in the next few hundred million years). Certainly not if by "nature" we mean the existence of life on Earth, the diversity of ancient lineages, or the ability of life to continue to evolve. Instead, what is under threat, what Bill McKibben announced the end of, are the life-forms that we most relate to and that are most integral to our own survival. What is threatened are the species we love and the species we need. This may seem like a case of semantics, but it isn't.

Nee's assertion really had two components. He pointed out what Figure 11.1 shows, our relative unimportance (and that of species like us) to the grandeur of life. In other words, he emphasized his support for the Erwinian revolution. But he was also noting that the conditions that we humans and other multicellular organisms prefer are a relatively narrow subset of those preferred by species more generally. Much of the biological world prefers conditions that are more extreme than the conditions we prefer or even those we can tolerate.

OUR OWN FAMILY on the tree of life, the hominids (which includes modern and extinct humans and modern and extinct apes), evolved roughly seventeen million years ago. By the time hominids began to evolve, essentially all the major branches on the tree of life had already been around for hundreds of millions or even billions of years. Some had lived through periods devoid of oxygen, others through periods with dangerously high concentrations of oxygen. Some had lived through extreme heat, others extreme cold. These lineages survived these changes as well as others (triggered by meteors, volcanoes, and more) either through broad tolerances or by finding small habitats, here and there, in which their preferred conditions, whatever they might be, persisted. The average conditions seventeen million years ago were relatively hostile for many lineages, but not for our own ancestors, the first hominids.

By the time the first monkey-sized hominids evolved, the oxygen levels in the environment were essentially those we now experience. Carbon dioxide levels were slightly higher though, as were temperatures. These were conditions that were conducive to early hominids. By the time *Homo erectus* evolved, about 1.9 million years ago, concentrations of oxygen and carbon dioxide and temperatures were essentially what we experience today, if anything a little cooler. They were conditions that we would now perceive as relatively pleasant. This isn't chance. Most of the features of our bodies related to our ability to withstand heat, the ability to sweat and even the details of our respiration, evolved during this period. Our lineage, in other words, like many modern lineages, is fine-tuned for the conditions of the last 1.9 million years, conditions that have been rare for nearly all of the long history of Earth.

Our bodies evolved to take advantage of a relatively unusual set of conditions that we think of as normal. It is easy to take those conditions for granted, but the truth is that the more we warm Earth, the less our bodies are suited to the world around us. The more we change the world, the more we increase the disconnect between the conditions we need to thrive and the world we live in. On the other hand, species that evolved their adaptations for temperature, gasses, and other conditions in the remote past and held on not through further adaptation but, instead, by finding small pockets of such conditions have the potential to persist and in some cases even to thrive, even as we make Earth warmer and, relative to our own needs and tolerances, polluted.

The conditions preferred by many ancient lineages of life include conditions that from our own perspective seem likely to be lifeless. Bacteria live at extraordinarily high pressures in volcanic vents at the bottom of the ocean and harvest energy from the core's hot exhaust. They have lived there for billions of years. One of those bacteria species, *Pyrolobus fumarri*, is the most thermally tolerant species on Earth. It can withstand temperatures of up to 112°C (235°F). Such bacteria die if brought to the surface, unable to deal with our pressures, unable to deal with sunlight, unable to deal with oxygen, unable to deal with the cold. Elsewhere, bacteria live inside salt crystals. They live in clouds. They live a mile underground growing on oil. The bacteria species *Deinococcus radiodurans* lives through radiation intense enough to weaken glass. The atomic bombs dropped on Hiroshima and Nagasaki in World War II contained one thousand rads of radiation. One thousand rads kills humans. *Deinococcus radiodurans* can withstand nearly two million rads. Nearly all (and perhaps all) of the extremes we are engendering on Earth correspond to at least some set of conditions from the past and, correspondingly, to some set of species capable of thriving. Any horror of the future is, to some species, a description of ideal conditions, especially if that future horror matches some period in the distant past.

However, we know very little about most of the species that will thrive in these new old conditions. With a few exceptions, ecologists haven't studied them. As I noted early in the book, ecologists have been overly focused on species like us, large-bodied, big-eyed mammal and bird species, many of which are very threatened by the changes we are causing. They have also been focused on the ecosystems and species that are declining rather than on those that might expand. Ecologists love to go to and study rain forests, ancient grasslands, and islands. They hate to work in toxic dumps and nuclear sites, even if the dumps and nuclear sites are proximate and relatively easy to study. And who can blame them? Meanwhile, the most extreme deserts on Earth are both remote and inhospitable, the kinds of places one is exiled to rather than the kinds of places to which one flocks when classes are done. They too are rarely studied. The result is that we tend to be blind to the ecology of some of the most rapidly growing ecosystems, blind to the future's extremes. Here, I'm no exception.

I became aware of this gap in our knowledge a few years ago when trying to understand how many and which ant species are likely to thrive in the context of climate change. One tool we used was a simple diagram called a Whittaker biome plot. The ecologist Robert Whittaker was in the habit of plotting temperature versus precipitation (a habit he appears to have picked up from the German and later German American ecologist Helmut Leith). He realized that these two variables on their own were enough to describe most of Earth's biomes. Hot and wet was a rain forest, hot and dry a desert, and so on. This relationship between climate and Earth's main biomes is sufficiently robust that the ecologist John Lawton called it one of "ecology's most useful generalisms."

A number of years ago, Nate Sanders (now a professor at the University of Michigan) and I coordinated a collaboration of tens of ant biologists from around the world. We worked together to pool all the studies we could find of ant communities from anywhere they had been systematically studied. Working with our colleague Clinton Jenkins, we then plotted the temperature and precipitation of the sites at which those studies were conducted. Each point represented hundreds of hours of work by some ant biologist. These data points were hard-won. Yet as we looked at the points relative to the climates found on Earth, we realized that something was missing.[5]

The places biologists had gone out to study ants were nonrandom with regard to climatic conditions. Some of the coldest conditions have not been studied, in part because many of those conditions lack ants. No one studies ants where they will not find ants. But the hottest forests and, especially, the hottest deserts have also been poorly studied. This isn't to say we know nothing about these sites, but our knowledge about these sites is particularly partial. We saw this pattern in ants, but it is almost certainly true for birds, mammals, plants, and most other groups of organisms. If we had considered other parameters, such as variability in temperature and precipitation or chemical features of the environment such as pH or salinity, we probably would have seen similar patterns. Generally speaking, it seems that the more extreme a set of conditions, from a human perspective, the less well the ant species living in those conditions are likely to have been studied.

YOU MIGHT ARGUE that the reason ant biologists haven't studied the communities of ants in the hottest deserts is that ants don't live there (in other words, something similar to what is going on in the coldest conditions). This isn't the case. Thanks to a relatively small number of heat-tolerant ant biologists, including my friend Xim Cerda, we know that some ant species, such as ants of the genus *Cataglyphis*, can take the heat. In fact, *Cataglyphis* ants tolerate temperatures hotter than any other animal species can survive. They forage in the hottest deserts in the world at the hottest times of day. *Cataglyphis* ants can survive temperatures up to 55°C (131°F), a full twenty-five degrees Celsius hotter than the hottest mean annual temperature of the air any place on Earth today. They are, as the entomologist Rüdiger Wehner put it, "heat-loving, heat-seeking, thermal warriors."[6] When it is hot, they collect flower petals. They lick sugars from plant stems. And they gather the dead bodies of other animal species that have died due to heat stress.

Cataglyphis ants have diversified in extreme habitats. No fewer than a hundred and potentially many more *Cataglyphis* species exist, each unique in its details and yet all fond of heat. These species have evolved a number of adaptations that allow them to cope with the heat. They have long legs that enable them to stay above the sand and run quickly, a flexible gaster (abdomen) that they can lift high above the sand, and bodies filled with heat-shock proteins that are constantly produced to help protect their cells and especially their enzymes from the temperatures to which they are exposed.[7] In addition, the most thermally tolerant *Cataglyphis* species, *Cataglyphis bombycina*, is covered with a dense layer of prism-like hairs that reflect nearly all visible and infrared light that lands on the ant; virtually none reaches the ant's body. This does not only keep them from warming; it also helps cool them by offloading some heat.[8]

The obvious challenge of studying these ants is that they prefer temperatures that are dangerously hot to other animals, including humans. Xim Cerda has studied these ants wherever he can find them. He has studied them in the hottest parts of Spain, in the Negev desert in Israel, on the dry Anatolian steppes in Turkey, and in the Sahara of Morocco. When he studies them, he must bring a lot of water. When that isn't enough, he sometimes buries himself in the sand to stay cool (Figure 11.2). Even so, there are days when the ants are active but he simply can't be, days when they thrive and his body fails. As Xim might say, this is partially because he isn't as young as he used to be, but it is also because he is human and, well, the ants are not. This is part of why there aren't many data points in the part of the Whittaker biome plot corresponding to high temperatures; those places are hard to study.

One of the places that *Cataglyphis* ants have yet to be studied, but almost certainly live, is in the Danakil desert in the northern part of the Afar Triangle of Ethiopia, along the borders of Eritrea and Djibouti. The Afar Triangle sits at the intersection of three continental plates, the Nubian plate, the Somali plate, and the Arabian plate. It is in the triangle that these plates are actively pulling away from each other, at around two centimeters a year. The Afar Triangle is a place of change; it used to be green. Grasslands and fig trees grew. In the region's rivers wandered hippos and swam giant catfish. On its hills giant hyenas ran after pigs,

Figure 11.2. When temperatures rise above what he can tolerate, Xim Cerda sometimes buries himself in the sand to study *Cataglyphis* ants (*left*). When temperatures get so hot that burying himself is not enough, Xim has other approaches to staying cool (*right*), though these latter approaches yield fewer data.

antelopes, and wildebeests. It used to look, more or less, like a small Serengeti. The ancient hominin *Ardipithecus ramidus* lived in the Afar Triangle 4.4 million years ago. The famous hominin Lucy and others of her species, *Australopithecus afarensis*, lived in the region around three to four million years ago. More recently, *Homo erectus* made stone tools in the region, hunted, and maybe even cooked. And *Homo sapiens*, our own species, was present in the region as early as 156,000 years ago. Throughout these millennia, the conditions in the region were conditions well within the limits of the ancient and modern human niches. Then drought arrived and stayed.

Today, the Danakil desert itself has few permanent inhabitants. Afar pastoralists bring their animals to feed in the area during the wetter season but then move on. The Danakil is a hard place to live. To European explorers, the challenges of even traveling through the region were akin to those posed by traveling in Antarctica, challenges of extremes. One chronicle describes a particularly hard journey through the desert during which "ten camels and three mules died of thirst, starvation and fatigue."[9] The conditions in this region are likely to become much more common in the coming years. Yet for as much as our ancestors once called the Afar Triangle home and paleoanthropologists spend many hours digging up their bones and stories, the modern ecology of the region is little known. No recent surveys of animal diversity appear to have been undertaken, nor have even the ant species known to live there been studied in any detail. Most studies of the animals of the region are studies of ancient, extinct, vertebrate species, studied on the basis of bones in fossils. That is a shame because the present conditions of the region, and especially the Danakil desert, are the most like those expected in many deserts in the future. The desert is extraordinarily hot, extraordinarily dry, and episodically but unpredictably flooded. It is almost certainly a land inhabited now by *Cataglyphis* species. But no one has yet studied these *Cataglyphis*, the species whose societies inherited the land once so fruitful to the societies of our ancestors; maybe one day Xim will (he has submitted one grant to do so, but the grant was not awarded by the funding agency). Maybe he won't.

The sands of the Danakil desert, on which *Cataglyphis* ants run, are a lens into the hostile climates likely to be more common in the future, a lens through which we have yet to look well. But they are not the most extreme habitat of the region. In one of the hottest, driest parts of the Danakil desert, one finds the Dallol geothermal area, its surface pocked with hot springs. The springs occur where sea water seeps beneath the land and makes contact with magma escaping from the center of the Earth. The water then rises up to the surface, producing hot springs akin to those found in Yellowstone National Park. The water is nearly 100°C (212°F) when it reaches the surface. It is also salty. Depending on the nature of the rock through which it rises, it is, in some places, also sulfurous or sulfurous and acidic. In some places, the pH of the water is 0. In very few other places on Earth have such acidic conditions been encountered. What is more, the air around the springs is so high in carbon dioxide that it kills animals that walk near them. Around the springs one finds the bones of birds and lizards that breathed in the carbon dioxide and suffocated or that mistook the springs for an oasis of sweet water and died due to their acidity. The air also contains, in some spots, potentially deadly concentrations of chlorine. The land in and around the springs is green, yellow, and white. It looks hostile. It smells hostile. It makes the desert that surrounds it, the hottest desert in the world, seem forgiving. Yet the springs are not hostile to all species. They actually abound in life.

Recently, Felipe Gómez from the Astrobiology Center in Spain and his colleagues discovered roughly a dozen species of Archaeans, the lineage discovered by Woese, that grow best in the hot, acidic, salty conditions of the springs. These dozen species are more evolutionarily diverse than all the vertebrates on Earth combined. These diverse, single-celled organisms may be the most extreme life-forms on Earth. They thrive in extreme conditions that are rare on Earth.[10] Gómez studies these species in part to understand the sorts of life-forms that might be found on other bodies in the solar system, such as Mars or Europa, the second moon of Jupiter. The microbes of the Dallol hot springs are the sorts of microbes that might be whisked by winds into the stratosphere and beyond and survive.[11] They might accidentally be carried by Mars rovers to the Red Planet. (They may already have been.) Or we might use them, in some way, to make life more suitable for us on Mars or elsewhere. But these microbes are also a measure of what life will be like in the harshest conditions we may inadvertently favor on Earth. These species wait for us to make Earth hotter, to make the soil saltier, and even to make conditions more acidic, so that they can thrive, so that more of Earth will, once again, be more hospitable to them.[12]

Conclusion

No Longer Among the Living

IN THE NEAR FUTURE, PARTS OF EARTH WILL BE MUCH MORE PLEASANT for extremophilic life-forms but much less suitable for humans. We can find ways to survive such change. Just not forever. Eventually, we will go extinct. All species do. This reality has been called the first law of paleontology.[1] The average longevity of animal species appears to be around two million years, at least for the taxonomic groups for which the phenomenon has been well studied.[2] If we consider just our species, *Homo sapiens*, that means we may still have some time. *Homo sapiens* evolved roughly two hundred thousand years ago. We are still a young species. This suggests that if we last an average amount of time, our road is still long. On the other hand, it is the youngest species that are most at risk of extinction. Like puppies, big-eyed and not yet wise, young species are prone to fatal mistakes.

The only species that tend to survive much longer than a few million years are microbes, some of which can go into long dormancy. Recently a research team in Japan gathered bacteria from deep beneath the sea. The bacteria were estimated to be more than a hundred million years old. The team gave the bacteria oxygen and food and then watched. After a few weeks the dormant bacteria, which had last respired during the dawn of mammals, began to respire again and divide.

It is tempting to imagine that in the far future, humans will figure out how to achieve bacteria-like suspended animation. But such imaginings are the sort of hubris to which our species has long been susceptible, the hubris of believing ourselves to be exempt from the laws of life. Our best bet for extending our stay on this planet is a humbler one: to pay attention to the laws of life and work with them rather than against them. We need to conserve and curate the islands of habitat on Earth in ways that promote the evolution of species that are benign or even beneficial to us. We need to provide corridors through which species can home in on the habitats in which they will be able to survive in future climates. We need to carefully manage the ecosystems around us so as to keep the parasites and pests of our bodies and crops at bay (to escape one more time). We need to reduce greenhouse gas emissions as fast as possible so as to leave as much of the Earth as possible with conditions that are still within the limits of the human niche. And we need to find ways to save the species and ecosystems on which we depend or might someday depend. And as we do all of this, we need to remember that we are one species among many, a species that is no more or less special than the shimmering hairy protist that lives inside the termite gut, the rhinoceros botfly, or a ground beetle species that lives its entire life among the leaves of a single individual of a single species of Panamanian tree.

We once assumed the sun circled Earth. Now we know that Earth circles the sun, which happens to be a rather mundane star, one among billions. We once assumed the story of life was about us. Now we know that the story of life is mostly about microbes. We are a clumsy giant late to the drama, a character in life's play that doesn't make it to the curtain call. We should, of course, try to prolong the time our species has on Earth, just as each of us tries to prolong our own life. Yet however much we extend what we have, it is good to be aware it is finite. We will end. When we end, the geological age defined by our consequences, the Anthropocene, will end. A new age will begin. We won't see it, and yet we can posit some of its features because, even after we are gone, species will continue to obey the laws of life.

THE FIRST THING we can predict about the future after us is which species are likely to miss us, or even to go extinct, when we're gone. The loss of species due to the extinction of the species on which they depend is called coextinction.

Years ago, in my collaboration with the Singaporean scientist Lian Pin Koh (now a member of parliament in Singapore), I wrote the first paper that tried to estimate how common such extinction might be in the world around us. Along with a team of clever collaborators, Lian Pin and I were concerned then about the dependent species that go extinct when rare plants and animals are lost. Because most species have other species that depend on them, coextinction is very common. We estimated that similar numbers of coextinctions have probably occurred as host extinctions. Many species have gone down with the corporeal ship, so to speak. However, the loss of dependent species is rarely very well documented because most dependent species are small-bodied and poorly studied, if studied at all.

Sometimes, dependent species go extinct even when their hosts just become rare, before they actually disappear entirely. When the black-footed ferret became very rare, dwindling to just a handful of individuals, it was brought into captivity to be reared and bred, and in the process, it was deloused. As a result of the dwindling size of its host's population, combined with that delousing, the black-footed ferret louse appears to have gone extinct. Subsequent attempts to find the black-footed ferret louse on ferrets have been unsuccessful.[3] The California condor's mites also appear to have been inadvertently extinguished when the condor was brought into captivity for breeding. Before these captive breeding programs, the black-footed ferret louse and the California condor's mites were coendangered. (Now they are coextinct.) Many thousands of species are now coendangered because of the rarity of the species on which they depend. The largest fly in Africa, the rhinoceros stomach botfly, lives only on the endangered black rhinoceros and the near-threatened white rhinoceros. Threats to the rhinos are threats to the fly.[4]

In studying coextinction and coendangerment, one of the things that Lian Pin and I learned is that two main factors determine how many species are threatened by the loss of a particular host. First, the more species a particular host supports, the more species will be coendangered if it becomes rare and the more species will suffer coextinctions due to its extinction. Second, the more specialized the dependent species on a particular host, the more likely they are to go extinct.

The classic example of a species on which many other specialized species depend and whose extinction would lead to many coextinctions is the army ant, *Eciton burchellii*. Army ants don't have a fixed home. They emigrate through the forest, eating what is before them, and then make a temporary home (a bivouac) out of their bodies, a palace of legs, abdomens, and heads. New colonies form when male army ants fly off, find another colony, and mate with new queens from that colony. The male then dies, and the fertilized queen forms her own new colony. To get a good start, she takes with her part of her mother's colony, the workers. The queen and workers leave together, on foot. Because of this form of colony founding, the species that live with army ants never need to fly or walk in search of a colony. They just need to follow either the old queen or the new one.

The unusual biology of army ants has led many species that depend on them to evolve; it has also led those dependent species to become very specialized. Dozens of species of mites live on the bodies of army ants. One of my favorites lives only on the mandibles of a single army ant species. Another lives only on army ant feet. Another still impersonates army ant larvae and lives among the larvae, where it is tended to as though it were a real army ant to be. Tens, and perhaps hundreds, of species of beetles ride the army ants from place to place or follow behind them. Silverfish tag along too, as do millipedes. Over millions of years, the number of species that live with and depend on each army ant species has grown and grown and grown.

Two of my mentors, Carl and Marian Rettenmeyer, spent their professional lives studying the species that live with army ants. They spent thousands and thousands of hours studying the species that live with army ants and little else. They traveled in search of these species. They dreamed of these species. Through such work, they estimated that the colonies of a single species of army ant, the aforementioned *Eciton burchellii*, host more than three hundred other species of animals (to say nothing of other life-forms, such as bacteria or viruses). Carl and Marian described the army ant species *Eciton burchellii* as the animal species on which the most other species depend. They called it the "largest animal association centered on one species."[5] And it appears to be, at least if one doesn't include humans.

During the great acceleration, a truly extraordinary diversity of species evolved to depend upon humans. The faster our human populations have grown, the faster they have been joined by ever larger numbers of dependent species, many of them, just as with army ants, quite specialized.

CONSIDER THE SPECIES that live with us. German cockroaches can survive nuclear radiation. Dust mites have survived in space (or at least one did, on the Russian *Mir* space station). Bedbugs are indefatigable. And, yes, Norway rats, black rats, and house mice have ridden with human colonists to nearly every island or continent. But these species survive best *with* us. They survive our assaults, assaults that kill off other species. In our absence, things are different.

In our absence, German cockroaches may well suffer co-extinction. Bedbugs will become as rare as they were before humans evolved, confined to bat caves and some bird nests. A version of this scenario became apparent in New York City during the period of peak quarantine in response to COVID-19. People moved out of Manhattan and also began spending less time outside. They spent less time out at dinner, less time eating on park benches, and less time out and about in general. As a result, less garbage accumulated, and the Norway rats of the city began to suffer. They became more aggressive. Their populations declined. The populations of other species dependent on human leftovers, such as pavement ants and house sparrows, are likely to have declined too.[6] Leftover-loving species need us.

But German cockroaches, bedbugs, and rats are just some of the most conspicuous of our dependents. It appears that more species depend on humans than have ever depended on any other species. Most primate species are host to tens of species of parasites; collectively, humans are host to thousands.[7] Our bodies also host beneficial gut bacteria, skin bacteria, vaginal bacteria, and oral bacteria that live nowhere else. These bacteria species, in turn, host unique viruses, bacteriophages, that rely on life that relies on us. Maybe there is some other contender for world's best host, but if so, I don't know what it is. The sum total of the number of body-dwelling species that will go extinct when we do is large. It is likely to include thousands, perhaps tens of thousands, of species.

Outside our bodies and homes, our dependents are even more diverse. In the time since the origin of agriculture, humans have domesticated hundreds of plant species and have bred and favored nearly a million different varieties among those species. Many of those varieties of crops are stored at the Svalbard Global Seed Vault in remotest Norway. However, the seed bank relies on humans in order to keep the seeds alive. Every so often, the seeds must be grown up to produce more seeds that can then be stored anew. Eventually, the seeds at Svalbard will all go extinct. In the big sweep of time, this will not take very long. By the time those varieties of seeds have gone extinct, the microbes that each of them relies on to grow will probably already be gone. Those microbes are not being conserved in Svalbard (except accidentally, inside some of the seeds). Instead, they are to be found only among crops in fields. With our loss, they will suffer coextinction, and so too will many of the specialist pests of our crops.

Some domestic animals will go extinct too. That includes cows and chickens. It may well include domestic dogs. Some dogs today are feral, yet feral dogs are rare outside human-populated areas. In most places, dogs need us to persist. The same is true of cats in some places, but not others. In Alaska, feral cat populations are short-lived. Those that aren't eaten don't make it through the winter. In Australia, on the other hand, hundreds of thousands of cats stalk the outback. The feral cats of Australia are likely to survive the extinction of human Australians. Goats will live on in many regions. With regard to the extinction of humans, goats are tougher than cockroaches.

The closest approximation to the sort of scenario in which the disappearance of humans precipitates the coextinction of other species occurred in Viking settlements in western Greenland. Vikings colonized Greenland beginning at the end of the tenth century CE. They farmed in several settlements while also hunting walruses, whose tusks they traded for goods that would otherwise be unavailable. Early on the Greenlandic Vikings lived in longhouses. Later they lived in more centrally planned settlements. In winter they kept their animals—including sheep, goats, cows, and a few horses—in stalls ringing their homes. Then, as climates cooled, their way of life collapsed, first in the northerly (and hence colder) western settlement and then later in the eastern settlement. Because this happened relatively recently, the years just after the collapse of the western settlement can be reconstructed based on archaeological studies and written records. Sometime before 1346, the inhabitants of at least two sites associated with the western settlement disappeared, having either fled or died. In 1346, Ívar Barðarsson visited one of those sites and found no humans. From archaeological studies, it is known that the common parasites of humans, specifically human lice and fleas, that had long been present at the site had also disappeared. Barðarsson did, however, find a few cows and sheep. And archaeological records also note from this period the parasites of sheep. Barðarsson ate some of the cows and left the other animals. Those animals may have survived a winter or two, but eventually their presence became undetectable at the site. As they disappeared, so did their parasites. Finally, the majority of the species left at the site were those with no relationship with humans whatsoever; they were the wildlife of Greenland, species going about their days as if the Vikings had never arrived.[8]

AFTER WE GO extinct, and after the last cow falls, life will be reborn from what is left. The species that remain will, as Alan Weisman put it in his book *The World Without Us*, "sigh a huge biological sigh of relief."[9] After the sigh, some features of the rebirth of Earth are predictable. The life that remains will be reshaped by natural selection into a diversity of new and wondrous forms. On some level, the details of those forms are unknowable, yet we do know that they will still obey life's laws.

If we consider the last half billion years of evolution, one of the clearest conclusions is that what comes after a mass extinction does not necessarily match up with what came before. The trilobites were not followed by more trilobites, nor were the largest herbivorous dinosaurs succeeded by more enormous dinosaurs, or even similarly sized mammalian herbivores (a cow is no brontosaurus). The details of the past do not necessarily predict those of the future (or vice versa). A version of this sentiment has been called the fifth law of paleontology.[10]

What can recur after mass extinctions are familiar themes, themes revisited by evolution the way one jazz musician might echo another jazz musician's riff. Evolutionary biologists call such themes *convergent*. They are cases in which two lineages, separated by space, history, or time, evolve similar features in light of similar conditions.

Sometimes convergent themes are subtle and idiosyncratic. The horns of the rhinoceros evoke the horns of *Triceratops*. In other cases, they are more obvious and grounded in the reality that often there are relatively few ways to live a particular lifestyle. Desert-dwelling lizards have evolved lacy toes, with which they can more easily run over the sand, half a dozen times. Ancient marine predators had sharklike shapes. Modern marine predators, including sharks but also dolphins and tuna, have nearly identical shapes. They also tend to have similar ways of moving (both mako sharks and tuna only move the last third of their body to swim). Ancient mammals that lived in burrows tended to have big butts (for blocking the burrow), at least one set of big feet for digging, and a propensity for storing food. So too do modern digging mammals that live similar lifestyles.

The extent of convergence in some lineages is startling; convergence can contain a sort of detail-rich sublimity. As the evolutionary biologist Jonathan Losos notes in his excellent book on convergent evolution, *Improbable Destinies*, African and American porcupines look very similar.[11] They have long spines. They waddle. They eat bark. They are, as mammals go, not extraordinarily clever. Yet they have evolved these attributes independently. They are no more closely related to each other than either is to a guinea pig. One episode of natural selection at a time, one generation at a time, they have stumbled toward an unusual and yet similar way of being.

In the white sand dunes of the Tularosa Basin in New Mexico, fence lizards and pocket mice have both evolved white coloration so as to stay hidden. Darker-colored lizards were seen by predators and eaten, their genes winnowed from the population one predation event at a time. In nearby tan-colored grasslands in the Tularosa Basin, their close relatives have tan and gray coloration to hide among the grasses. And out in the lava fields of the Tularosa Basin, others of their relatives have evolved near-black coloration to match the lava stone.[12] What are the limits of this change? Were we to paint the desert pink, would the lizards evolve pinkness? Yellowness? Maybe. If they had the right genetic variation; if they had enough time.

Elsewhere, in dry deserts, small mammals have half a dozen times evolved a tendency to hop on two legs. In hot, dry deserts featuring plants that accumulate salt, mammals have at least twice evolved hairs in their mouths to strip that salt from the plant (so as to be able to eat its leaves) and kidneys especially well tuned to deal with high salt. Meanwhile, on islands, big mammals tend to evolve smaller sizes (miniature elephants, miniature mammoths). Smaller animals, in the absence of bigger animals, evolve bigger sizes (giant ground-dwelling Caribbean owls). Similarly, as I noted earlier, flighted animals evolve flightlessness. A recent study concluded that flightlessness among birds evolved far more often on islands than was suspected, more than a hundred times. We had missed this fact; we had overlooked the stubby-winged waddle-beasts of many archipelagos. They were easy to miss because these birds were the most susceptible to extinction once humans arrived. By the time humans started to tally the living world, they were already gone.[13]

In some cases, our understanding of the examples of convergent evolution is detailed, formalized in the rigors of experiment, mathematics, and data. Jonathan Losos has spent his career studying the anole lizards of the Caribbean. Like a witch's cauldron, his mind is populated with lizard tails and feet. Through careful study, Losos has been able to show that when anole lizards arrive on Caribbean islands they predictably, one might even say inevitably, evolve into three basic forms. Some evolve into species that live in the canopy with hairy feet suited for hanging on to branches and twigs. Others evolve to live on twigs. These also have hairy feet, but they have short legs and short tails that enable them to keep from falling from the twigs. And some evolve into ground runners with long legs and small toe pads. These forms have evolved independently one or more times on each of the four large islands of the Caribbean. There are only so many ways, it appears, to succeed as a Caribbean anole.[14]

Then, of course, there are the kinds of convergence I've already discussed in this book, the convergence that occurs, rapidly, in the face of the deadly force that humanity presses upon nature. Resistant bacteria, insects, weeds, and fungi evolve predictably. Often their resistance is due to convergent traits. The repeatability of the megaplate experiment that Michael Baym orchestrated was due to convergence. In some cases, the convergence relates not just to the evolution of resistance or to the mechanisms by which that resistance protects the species we attack, but even to the genes that enable such resistance.

THE MANY EXAMPLES of convergent evolution suggest the rules of life that will influence what kinds of species will evolve anew in the future. In general, these examples suggest general tendencies of evolution rather than details of the biology of individual species. Yet sometimes predictions have been successful in the past even when considering details of species. For example, Richard Alexander, a faculty member at the University of Michigan, long studied the evolution of insect societies such as those of ants, bees, termites, and wasps. In all of these societies, some individuals (the queens and the kings) reproduce, but most do not. These nonreproductive individuals, called workers, work on behalf of their queen and king. Such societies are known as eusocial societies.

Eusocial societies are especially unusual in an evolutionary sense. In evolution, the only "goal" organisms have is to pass on their own genes, and yet among ants, bees, termites, and wasps workers forgo their chance. Workers care for eggs and babies. They gather food. They defend the colony. But they do not, except in exceptional cases, reproduce.

The sole evolutionary benefit to workers of forgoing reproduction is that, in doing so, they help improve the success of their relatives' genes; a relatively high proportion of which are the same as their own. Alexander identified a set of circumstances in which eusocial societies with such workers should evolve. Eusocial societies, Alexander noted, tend to convergently evolve when individuals living together are close relatives and hence have similar genes. They tend to evolve when food is patchy (and the patches are enough to support more than one individual). And they tend to evolve in circumstances in which individuals working together can readily defend their home. At least this was the case for insects. For example, as noted earlier in the book, termites evolved from cockroaches in the context of the confined space of logs. In such logs, inbreeding is thought to have been common (and hence individuals were closely related), and the food and home were one and the same and were patchy and defensible.

There are no eusocial birds, reptiles, or amphibians, and at the time Alexander was writing, no truly eusocial mammals had been discovered. However, in a series of lectures at North Carolina State University and elsewhere, beginning in 1975, Alexander predicted that such mammals might exist. Alexander wasn't forecasting the future. Instead, he was trying to anticipate the yet-to-be-studied details of the contemporary world. Alexander offered twelve detailed predictions about the biology of the yet-to-be-discovered mammal.[15] The mammal would live in a seasonal desert. It would live underground and subsist on roots. It would also probably be a rodent. Alexander announced his predictions at talk after talk. Finally, in 1976 while he was giving the talk again, this time at Northern Arizona University, an audience member, the mammalogist Richard Vaughan, stood up and said something to the effect of, "Well, excuse me, but that sounds like a description of a naked mole rat." Subsequent research by the mammalogist Jennifer Jarvis would reveal naked mole rats to be the embodiment of Alexander's predictions: eusocial mammals living underground in the desert, naked but for their sagging skin, eating roots.[16]

It would be interesting to gather a group of evolutionary biologists and ask them what Alexander-like predictions they would offer with regard to life after us. My informal survey of my colleagues suggests that they would tend to agree that the way the evolution of new species proceeds in our absence depends on how much is lost. In general, though, they would also agree that life tends to become more diverse, varied, and complex over time, a sentiment that is also sometimes considered a law of paleontology. So if there is a species of a lineage left, and it survives, it will become more than one species. Considering mammals: if there are still representatives of the major groups of mammals, they might evolve anew in the ways they evolved in the past. If there are half a dozen species of wild cats left, each might, depending on its location and details, evolve into a dozen different new cat species, some bigger, some smaller. The same with canids: from one species of wolf or fox, many new species. Some species might be remarkably similar to those we are familiar with today; others will be unpredictably different. We actually have evidence of something much like this from the past. Predatory mammals evolved both among placental mammals and among marsupial mammals. The gray wolf is a placental mammal; the thylacine was a predatory marsupial mammal. Recently, Christy Hipsley, an assistant professor at the University of Copenhagen, compared in great detail the skulls of a sampling of placental mammals and a sampling of marsupial mammals. She found that the skull of the thylacine was more similar to the skull of the gray wolf than to that of any marsupial species yet studied. The two species show remarkable convergence on a predictably good way to be a medium-sized carnivore. On the other hand, many marsupial mammals, including the wombat, resembled other marsupial mammals far more than they did any placental mammal.[17]

The colleagues I surveyed, including Jonathan Losos, also agreed about one other predictable feature of the rediversification of cats, or any other group of mammals. In general, when conditions are colder, warm-blooded animals tend to evolve larger body sizes. Larger-bodied animals have proportionally less surface area over which to lose heat. Conversely, when temperatures are warmer, they tend to evolve smaller body sizes (this is called Bergmann's rule or law). Smaller-bodied animals have more surface area across which they can sweat or otherwise lose heat. If humans go extinct far in the future during a glacial cycle, larger-bodied individuals may be more likely to survive, and hence larger bodies may evolve in many lineages.

If we disappear during warmer times, many species, particularly mammal species, may evolve smaller body sizes. The evolution of small-bodied mammals is well documented during the last period in which Earth was extraordinarily hot. Tiny horses evolved.[18] Natural selection has no sense of whimsy. It has no sense of anything, and yet the reality that tiny horses once existed, prancing about in the ancient warmth, is as whimsical a thing as I can imagine. The effects of heat on body size can also be seen in the recent past by considering individual species. Over the last twenty-five thousand years, the body size of woodrats in the desert southwest has tracked changes in climate. When it was hot, their bodies shrank. When it was cooler, they became larger.[19]

If we leave in our wake a wave of more extreme extinctions, natural selection might more actively reinvent the world, futzing with the leftover pieces and bits at its disposal. Imagining a scenario in which most mammal species have gone extinct, the authors of *The Earth After Us*, Jan Zalasiewicz and Kim Freedman, posited a whole suite of new kinds of mammals that might evolve.[20] They started with the assumption that the organisms most likely to diversify would be those that are already widespread, could live without humans, and would be isolated by our absence (which would also mean the absence of boats, planes, cars, and other sources of transport). They thought rats meet these criteria; rats would be the future. Some rat species and populations are very dependent on humans (and hence our existence). However, there are many rat species and even some populations of human-associated rat species that are not; these might then beget the future mammal fauna. If they do, one might, Zalasiewicz and Freedman wrote,

> imagine, perhaps, a diversity of rodents derived from our present-day rats. . . . Their descendants may be of various shapes and sizes; some smaller than shrews, and others the size of elephants, roaming the grasslands, yet others are swift and strong and deadly as leopards. We might include among them—for curiosity's sake and to keep our options open—a species or two of large naked rodent, living in caves, shaping rocks as primitive tools and wearing the skins of other mammals that they have killed and eaten. In the oceans, we might envisage seal-like rodents and, hunting them, ferocious killer

> rodents, sleek and streamlined as the dolphins of today and the ichthyosaurs of yesteryear.[21]

In addition to the evolutionary scenarios we can imagine, whether in light of life's convergent tendencies or other processes, it is tempting to ponder those that might be so different that they are not anticipated anywhere in the life we know. Could we really imagine elephants if they did not exist? Or woodpeckers? Their unique lifestyles and features (trunk and wood-pecking beak, respectively) have evolved just once. But I suspect we are not creative enough to imagine the species that both might be favored by evolution and are truly different from those we know. When painters try to imagine such species, they often give animals extra heads (Alexis Rockman) or legs (Rockman again, though also Hieronymus Bosch). Or they combine traits of different organisms into one (saber teeth, deer antlers, rabbit ears, and cloven hooves). The results tend to seem either too much a hodgepodge to be viable (the multiple heads) or too unusual to be probable. Yet if we are honest, so too do some of the species we find around us on Earth. The platypus, for instance, has a duck's beak, webbed feet, poisonous spurs, and an assortment of other oddities. Could we imagine a platypus if we didn't know it to exist?

It is common, in pondering the unusual features of the far future, to consider whether any of these species that succeed us might evolve the sort of intelligence that we find impressive, which is to say intelligence like our own (the kind that leads a species to warm its planet to its own detriment). Could the future after us be a future of ever-cleverer crows or, say, city-building dolphins? The unambiguous answer is maybe. I asked Jonathan Losos about the future of intelligent life in an interview, and he thought that, given enough time, some other primate might evolve humanlike intelligence. Maybe. But if we extinguish primates, he was less sure.[22] And anyway, the sort of intelligence we know so far on Earth is only helpful in a subset of situations. It is useful when conditions are uncertain from year to year. Yet even this has some upper bound; there is some level of uncertainty beyond which a big brain is no longer helpful. Perhaps that is just what will ultimately befall us, an Earth on which we have engendered conditions that are just too unpredictable year to year for us to solve through our inventive intelligence. Sometimes conditions can be so challenging that the species that survive are not the smart ones but, instead, the lucky ones and the fecund ones. In the contest between the clever crow and the fecund pigeon, sometimes the pigeon wins.

Then again, maybe a different sort of inventive intelligence will thrive anew in the future. A number of recent books have recently reconsidered, with some urgency, whether some sort of artificial intelligence, distributed among different machines, might take over Earth. These machines would be able to learn and would replicate, somewhere, out in the wild. Can we be on the road to creating artificially intelligent computer systems that might reproduce themselves after we are gone? They would need to find energy. They would need to be able to repair themselves. Yet there are plenty of books about just this possibility. I'll leave it to those books to ponder whether the computers—roving, thinking, mating, self-sustaining computers—take over. Meanwhile, it is interesting that in some ways we find it easier to posit that we can invent another entity that can live sustainably than to imagine that we can do so ourselves.

But there is also another kind of intelligence, distributed intelligence, the sort found in honey bees, termites, and, especially, ants. Ants are not inventively intelligent, at least not individually. Instead, their intelligence stems from their ability to apply rules about how to deal with new circumstances. Those fixed rules allow creativity to emerge in the form of collective behaviors. Looked at this way, ants and other societies of insects were computers before computers. Their intelligence is different from our own. They are not self-aware. They don't anticipate the future. They don't mourn the loss of other species, or even their own dead. Yet they can build structures that last. The oldest termite mound may well have been inhabited for longer than the oldest human city. Social insects can farm sustainably. Leaf-cutter ants farm fungi on fresh leaves, fungi that they then feed to their babies. Leaf-cutting termites do the same on dead leaves. They can make bridges of their bodies. They are everything that one imagines self-teaching robots might someday be, with the additional features that they are alive, they already exist, and they control a proportion of Earth's biomass roughly as large as the amount we control. They run their worlds more quietly than we run our own, and yet collectively, they run them all the same. In our absence, they would thrive as rulers, at least for a while, until they too went extinct.

After the societies of insects, the world is likely to be microbial, as it was for so long in the beginning and, if we are honest, as it has always been. As the paleontologist Stephen Jay Gould put it in his book *Full House*, "Our planet has always been in the 'Age of Bacteria,' ever since the first fossils—bacteria, of course—were entombed in rocks."[23] Once the ants are gone, it will remain the age of bacteria, or more generally microbial life, at least until conditions eventually become, for any of a variety of cosmic reasons, too extreme for microbes too. Then it will be quiet, a planet, once more, moved by physics and chemistry alone, a planet on which the innumerable rules of life no longer apply.[24]

Notes

Victoria Pryor, T. J. Kelleher, and Brandon Proia provided useful editorial suggestions throughout the book. Christa Clapp helped me think of the consequences of ecology's laws from an investor's perspective. The Department of Applied Ecology at North Carolina State University and the Center for Evolutionary Hologenomics at the University of Copenhagen provided the context in which much of the work this book builds on took place. The National Science Foundation funded much of the research on which the insights of this book are built; through basic biology we understand the general truths that allow practical action. This book would not have been possible without the generous support of the Sloan Foundation. I'm especially grateful to Doron Weber, who saw what this book could become (and hopefully became). As always, the biggest thanks are owed to Monica Sanchez, who had to listen to me talk about life's laws when I woke up at 2 a.m. with an idea, who ate more than one breakfast while listening to me talk about the geography of disease, and who walked along picturesque Danish coastlines while discussing sea-level rise. Thank you, Monica.

Introduction

1. Ghosh, Amitav, *The Great Derangement: Climate Change and the Unthinkable* (Chicago University Press, 2016), 5.
2. Ammons, A. R., "Downstream," in *Brink Road* (W. W. Norton, 1997).
3. Weiner, J., *The Beak of the Finch: A Story of Evolution in Our Time* (Knopf, 1994), 298.

4. Martin Doyle provided very useful insights about the Mississippi and its workings. See Martin's extraordinary book about America's rivers: Doyle, Martin, *The Source: How Rivers Made America and America Remade Its Rivers* (W. W. Norton, 2018).

Chapter 1: Blindsided by Life

1. Steffen, W., W. Broadgate, L. Deutsch, O. Gaffney, and C. Ludwig, "The Trajectory of the Anthropocene: The Great Acceleration," *Anthropocene Review* 2, no. 1 (2015): 81–98.

2. Comte de Buffon, Georges-Louis Leclerc, *Histoire naturelle, générale et particulière*, vol. 12, *Contenant les époques de la nature* (De L'Imprimerie royale, 1778).

3. Gaston, Kevin J., and Tim M. Blackburn, "Are Newly Described Bird Species Small-Bodied?," *Biodiversity Letters* 2, no. 1 (1994): 16–20.

4. National Research Council, *Research Priorities in Tropical Biology* (US National Academy of Sciences, 1980).

5. Rice, Marlin E., "Terry L. Erwin: She Had a Black Eye and in Her Arm She Held a Skunk," *ZooKeys* 500 (2015): 9–24; originally published in *American Entomologist* 61, no. 1 (2015): 9–15.

6. Erwin, Terry L., "Tropical Forests: Their Richness in Coleoptera and Other Arthropod Species," *The Coleopterists Bulletin* 36, no. 1 (1982): 74–75.

7. Stork, Nigel E., "How Many Species of Insects and Other Terrestrial Arthropods Are There on Earth?," *Annual Review of Entomology* 63 (2018): 31–45.

8. Barberán, Albert, et al., "The Ecology of Microscopic Life in Household Dust," *Proceedings of the Royal Society B: Biological Sciences* 282, no. 1814 (2015): 20151139.

9. Locey, Kenneth J., and Jay T. Lennon, "Scaling Laws Predict Global Microbial Diversity," *Proceedings of the National Academy of Sciences* 113, no. 21 (2016): 5970–5975.

10. Erwin, quoted in Strain, Daniel, "8.7 Million: A New Estimate for All the Complex Species on Earth," *Science* 333, no. 6046 (2011): 1083.

11. The origin of this quotation is described in Robinson, Andrew, "Did Einstein Really Say That?," *Nature* 557, no. 7703 (2018): 30–31.

12. Liu, Li, Jiajing Wang, Danny Rosenberg, Hao Zhao, György Lengyel, and Dani Nadel, "Fermented Beverage and Food Storage in

13,000 Y-Old Stone Mortars at Raqefet Cave, Israel: Investigating Natufian Ritual Feasting," *Journal of Archaeological Science: Reports* 21 (2018): 783–793.

13. Based on estimates by Jack Longino.

14. Hallmann, Caspar A., et al., "More Than 75 Percent Decline over 27 Years in Total Flying Insect Biomass in Protected Areas," *PLOS ONE* 12, no. 10 (2017): e0185809.

15. Thanks to Brian Wiegmann, Michelle Trautwein, Frido Welker, Martin Doyle, Nigel Stork, Ken Locey, Jay Lennon, Karen Lloyd, and Peter Raven for reading and thoughtfully commenting on this chapter. Thomas Pape provided especially generous and useful comments.

Chapter 2: Urban Galapagos

1. Wilson, Edward O., *Naturalist* (Island Press, 2006), 15.

2. Gotelli, Nicholas J., *A Primer of Ecology*, 3rd ed. (Sinauer Associates, 2001), 156.

3. Moore, Norman W., and Max D. Hooper, "On the Number of Bird Species in British Woods," *Biological Conservation* 8, no. 4 (1975): 239–250.

4. Williams, Terry Tempest, *Erosion: Essays of Undoing* (Sarah Crichton Books, 2019), ix.

5. Quammen, David, *The Song of the Dodo: Island Biogeography in an Age of Extinction* (Scribner, 1996); Kolbert, Elizabeth, *The Sixth Extinction: An Unnatural History* (Henry Holt, 2014).

6. Chase, Jonathan M., Shane A. Blowes, Tiffany M. Knight, Katharina Gerstner, and Felix May, "Ecosystem Decay Exacerbates Biodiversity Loss with Habitat Loss," *Nature* 584, no. 7820 (2020): 238–243.

7. MacArthur, R. H., and E. O. Wilson, *The Theory of Island Biogeography*, Princeton Landmarks in Biology (Princeton University Press, 2001), 152.

8. Darwin, Charles, *Journal of Researches into the Geology and Natural History of the Various Countries Visited by H.M.S. Beagle, Under the Command of Captain FitzRoy, R.N., from 1832 to 1836* (Henry Colborun, 1839), in chap. 17.

9. Coyne, Jerry A., and Trevor D. Price, "Little Evidence for Sympatric Speciation in Island Birds," *Evolution* 54, no. 6 (2000): 2166–2171.

10. Darwin, Charles, *On the Origin of Species*, 6th ed. (John Murray, 1872), in chap. 13.

11. Quammen, *The Song of the Dodo*, 19.

12. Izzo, Victor M., Yolanda H. Chen, Sean D. Schoville, Cong Wang, and David J. Hawthorne, "Origin of Pest Lineages of the Colorado Potato Beetle (Coleoptera: Chrysomelidae)," *Journal of Economic Entomology* 111, no. 2 (2018): 868–878.

13. Martin, Michael D., Filipe G. Vieira, Simon Y. W. Ho, Nathan Wales, Mikkel Schubert, Andaine Seguin-Orlando, Jean B. Ristaino, and M. Thomas P. Gilbert, "Genomic Characterization of a South American Phytophthora Hybrid Mandates Reassessment of the Geographic Origins of *Phytophthora infestans*," *Molecular Biology and Evolution* 33, no. 2 (2016): 478–491.

14. McDonald, Bruce A., and Eva H. Stukenbrock, "Rapid Emergence of Pathogens in Agro-Ecosystems: Global Threats to Agricultural Sustainability and Food Security," *Philosophical Transactions of the Royal Society B: Biological Sciences* 371, no. 1709 (2016): 20160026.

15. Puckett, Emily E., Emma Sherratt, Matthew Combs, Elizabeth J. Carlen, William Harcourt-Smith, and Jason Munshi-South, "Variation in Brown Rat Cranial Shape Shows Directional Selection over 120 Years in New York City," *Ecology and Evolution* 10, no. 11 (2020): 4739–4748.

16. Combs, Matthew, Kaylee A. Byers, Bruno M. Ghersi, Michael J. Blum, Adalgisa Caccone, Federico Costa, Chelsea G. Himsworth, Jonathan L. Richardson, and Jason Munshi-South, "Urban Rat Races: Spatial Population Genomics of Brown Rats (*Rattus norvegicus*) Compared Across Multiple Cities," *Proceedings of the Royal Society B: Biological Sciences* 285, no. 1880 (2018): 20180245.

17. Cheptou, P.-O., O. Carrue, S. Rouifed, and A. Cantarel, "Rapid Evolution of Seed Dispersal in an Urban Environment in the Weed *Crepis sancta*," *Proceedings of the National Academy of Sciences* 105, no. 10 (2008): 3796–3799.

18. Thompson, Ken A., Loren H. Rieseberg, and Dolph Schluter, "Speciation and the City," *Trends in Ecology and Evolution* 33, no. 11 (2018): 815–826.

19. Palopoli, Michael F., Daniel J. Fergus, Samuel Minot, Dorothy T. Pei, W. Brian Simison, Iria Fernandez-Silva, Megan S. Thoemmes, Robert R. Dunn, and Michelle Trautwein, "Global Divergence of the Human Follicle Mite *Demodex folliculorum*: Persistent Associations Between Host

Ancestry and Mite Lineages," *Proceedings of the National Academy of Sciences* 112, no. 52 (2015): 15958–15963.

20. I am very grateful to Christina Cowger, Fred Gould, Jean Ristaino, Yael Kisel, Tim Barraclough, Jason Munshi-South, Ryan Martin, Nate Sanders, Will Kimler, George Hess, and Nick Gotelli, all of whom provided useful comments on this chapter.

Chapter 3: The Inadvertent Ark

1. Pocheville, Arnaud, "The Ecological Niche: History and Recent Controversies," in *Handbook of Evolutionary Thinking in the Sciences*, ed. Thomas Heams, Philippe Huneman, Guillaume Lecointre, and Marc Silberstein (Springer, 2015), 547–586.

2. Munshi-South, Jason, "Urban Landscape Genetics: Canopy Cover Predicts Gene Flow Between White-Footed Mouse (*Peromyscus leucopus*) Populations in New York City," *Molecular Ecology* 21, no. 6 (2012): 1360–1378.

3. Finkel, Irving, *The Ark Before Noah: Decoding the Story of the Flood* (Hachette UK, 2014).

4. Terando, Adam J., Jennifer Costanza, Curtis Belyea, Robert R. Dunn, Alexa McKerrow, and Jaime A. Collazo, "The Southern Megalopolis: Using the Past to Predict the Future of Urban Sprawl in the Southeast US," *PLOS ONE* 9, no. 7 (2014): e102261.

5. Kingsland, Sharon E., "Urban Ecological Science in America," in *Science for the Sustainable City: Empirical Insights from the Baltimore School of Urban Ecology*, ed. Steward T. A. Pickett, Mary L. Cadenasso, J. Morgan Grove, Elena G. Irwin, Emma J. Rosi, and Christopher M. Swan (Yale University Press, 2019), 24.

6. Carlen, Elizabeth, and Jason Munshi-South, "Widespread Genetic Connectivity of Feral Pigeons Across the Northeastern Megacity," *Evolutionary Applications* 14, no. 1 (2020): 150–162.

7. Tang, Qian, Hong Jiang, Yangsheng Li, Thomas Bourguignon, and Theodore Alfred Evans, "Population Structure of the German Cockroach, *Blattella germanica*, Shows Two Expansions Across China," *Biological Invasions* 18, no. 8 (2016): 2391–2402.

8. Thanks to Adam Terando, George Hess, Nate Sanders, Nick Haddad, Jen Costanza, Jason Munshi-South, Doug Levey, Heather Cayton, and Curtis Belyea, all of whom read this chapter and provided helpful comments.

Chapter 4: The Last Escape

1. Xu, Meng, Xidong Mu, Shuang Zhang, Jaimie T. A. Dick, Bingtao Zhu, Dangen Gu, Yexin Yang, Du Luo, and Yinchang Hu, "A Global Analysis of Enemy Release and Its Variation with Latitude," *Global Ecology and Biogeography* 30, no. 1 (2021): 277–288.

2. Seyfarth, Robert M., Dorothy L. Cheney, and Peter Marler, "Monkey Responses to Three Different Alarm Calls: Evidence of Predator Classification and Semantic Communication," *Science* 210, no. 4471 (1980): 801–803.

3. Headland, Thomas N., and Harry W. Greene, "Hunter-Gatherers and Other Primates as Prey, Predators, and Competitors of Snakes," *Proceedings of the National Academy of Sciences* 108, no. 52 (2011): E1470–E1474.

4. Dunn, Robert R., T. Jonathan Davies, Nyeema C. Harris, and Michael C. Gavin, "Global Drivers of Human Pathogen Richness and Prevalence," *Proceedings of the Royal Society B: Biological Sciences* 277, no. 1694 (2010): 2587–2595.

5. Varki, Ajit, and Pascal Gagneux, "Human-Specific Evolution of Sialic Acid Targets: Explaining the Malignant Malaria Mystery?," *Proceedings of the National Academy of Sciences* 106, no. 35 (2009): 14739–14740.

6. Loy, Dorothy E., Weimin Liu, Yingying Li, Gerald H. Learn, Lindsey J. Plenderleith, Sesh A. Sundararaman, Paul M. Sharp, and Beatrice H. Hahn, "Out of Africa: Origins and Evolution of the Human Malaria Parasites *Plasmodium falciparum* and *Plasmodium vivax*," *International Journal for Parasitology* 47, nos. 2–3 (2017): 87–97.

7. For more on the story of the evolution of these parasites, see Kidgell, Claire, Ulrike Reichard, John Wain, Bodo Linz, Mia Torpdahl, Gordon Dougan, and Mark Achtman, "*Salmonella typhi*, the Causative Agent of Typhoid Fever, Is Approximately 50,000 Years Old," *Infection, Genetics and Evolution* 2, no. 1 (2002): 39–45.

8. Araújo, Adauto, and Karl Reinhard, "Mummies, Parasites, and Pathoecology in the Ancient Americas," in *The Handbook of Mummy Studies: New Frontiers in Scientific and Cultural Perspectives*, ed. Dong Hoon Shin and Raffaella Bianucci (Springer, forthcoming).

9. Bos, Kirsten I., et al., "Pre-Columbian Mycobacterial Genomes Reveal Seals as a Source of New World Human Tuberculosis," *Nature* 514, no. 7523 (2014): 494–497.

10. Wolfe, Nathan D., Claire Panosian Dunavan, and Jared Diamond, "Origins of Major Human Infectious Diseases," *Nature* 447, no. 7142 (2007): 279–283.

11. Koch, Alexander, Chris Brierley, Mark M. Maslin, and Simon L. Lewis, "Earth System Impacts of the European Arrival and Great Dying in the Americas After 1492," *Quaternary Science Reviews* 207 (2019): 13–36.

12. Matile-Ferrero, D., "Cassava Mealybug in the People's Republic of Congo," in *Proceedings of the International Workshop on the Cassava Mealybug Phenacoccus manihoti Mat.-Ferr. (Pseudococcidae)*, held at INERA-M'vuazi, Bas-Zaire, Zaire, June 26–29, 1977 (International Institute of Tropical Agriculture, 1978), 29–46.

13. Cox, Jennifer M., and D. J. Williams, "An Account of Cassava Mealybugs (Hemiptera: Pseudococcidae) with a Description of a New Species," *Bulletin of Entomological Research* 71, no. 2 (1981): 247–258.

14. Bellotti, Anthony C., Jesus A. Reyes, and Ana María Varela, "Observations on Cassava Mealybugs in the Americas: Their Biology, Ecology and Natural Enemies," in Sixth Symposium of the International Society for Tropical Root Crops, 339–352 (1983).

15. Herren, H. R., and P. Neuenschwander, "Biological Control of Cassava Pests in Africa," *Annual Revue of Entomology* 36 (1991): 257–283.

16. I tell the story of the cassava mealybug in more detail in Dunn, Rob, *Never Out of Season: How Having the Food We Want When We Want It Threatens Our Food Supply and Our Future* (Little, Brown, 2017).

17. Onokpise, Oghenekome, and Clifford Louime, "The Potential of the South American Leaf Blight as a Biological Agent," *Sustainability* 4, no. 11 (2012): 3151–3157.

18. Stensgaard, Anna-Sofie, Robert R. Dunn, Birgitte J. Vennervald, and Carsten Rahbek, "The Neglected Geography of Human Pathogens and Diseases," *Nature Ecology and Evolution* 1, no. 7 (2017): 1–2.

19. Fitzpatrick, Matt, "Future Urban Climates: What Will Cities Feel Like in 60 Years?," University of Maryland Center for Environmental Science, www.umces.edu/futureurbanclimates.

20. Thanks to Hans Herren, Jean Ristaino, Ainara Sistiaga Gutiérrez, Ajit Varki, Charlie Nunn, Matt Fitzpatrick, Anna-Sofie Stensgaard, Beatrice Hahn, Beth Archie, and Michael Reiskind for reading and commenting on versions of this chapter.

Chapter 5: The Human Niche

1. Xu, Chi, Timothy A. Kohler, Timothy M. Lenton, Jens-Christian Svenning, and Marten Scheffer, "Future of the Human Climate Niche," *Proceedings of the National Academy of Sciences* 117, no. 21 (2020): 11350–11355.

2. Manning, Katie, and Adrian Timpson, "The Demographic Response to Holocene Climate Change in the Sahara," *Quaternary Science Reviews* 101 (2014): 28–35.

3. Hsiang, Solomon M., Marshall Burke, and Edward Miguel, "Quantifying the Influence of Climate on Human Conflict," *Science* 341, no. 6151 (2013), https://doi.org/10.1126/science.1235467.

4. Larrick, Richard P., Thomas A. Timmerman, Andrew M. Carton, and Jason Abrevaya, "Temper, Temperature, and Temptation: Heat-Related Retaliation in Baseball," *Psychological Science* 22, no. 4 (2011): 423–428.

5. Kenrick, Douglas T., and Steven W. MacFarlane, "Ambient Temperature and Horn Honking: A Field Study of the Heat/Aggression Relationship," *Environment and Behavior* 18, no. 2 (1986): 179–191.

6. Rohles, Frederick H., "Environmental Psychology—Bucket of Worms," *Psychology Today* 1, no. 2 (1967): 54–63.

7. Almås, Ingvild, Maximilian Auffhammer, Tessa Bold, Ian Bolliger, Aluma Dembo, Solomon M. Hsiang, Shuhei Kitamura, Edward Miguel, and Robert Pickmans, *Destructive Behavior, Judgment, and Economic Decision-Making Under Thermal Stress*, working paper 25785 (National Bureau of Economic Research, 2019), https://www.nber.org/papers/w25785.

8. Burke, Marshall, Solomon M. Hsiang, and Edward Miguel, "Global Non-Linear Effect of Temperature on Economic Production," *Nature* 527, no. 7577 (2015): 235–239.

9. Thanks to Solomon Hsiang, Mike Gavin, Jens-Christian Svenning, Chi Xu, Matt Fitzpatrick, Nate Sanders, Edward Miguel, Ingvild Almås, and Maarten Scheffer, who read this chapter and provided thoughtful comments.

Chapter 6: The Intelligence of Crows

1. Pendergrass, Angeline G., Reto Knutti, Flavio Lehner, Clara Deser, and Benjamin M. Sanderson, "Precipitation Variability Increases in a Warmer Climate," *Scientific Reports* 7, no. 1 (2017): 1–9; Bathiany, Sebastian, Vasilis Dakos, Marten Scheffer, and Timothy M. Lenton, "Climate

Models Predict Increasing Temperature Variability in Poor Countries," *Science Advances* 4, no. 5 (2018): eaar5809.

2. Diamond, Sarah E., Lacy Chick, Abe Perez, Stephanie A. Strickler, and Ryan A. Martin, "Rapid Evolution of Ant Thermal Tolerance Across an Urban-Rural Temperature Cline," *Biological Journal of the Linnean Society* 121, no. 2 (2017): 248–257.

3. Grant, Barbara Rosemary, and Peter Raymond Grant, "Evolution of Darwin's Finches Caused by a Rare Climatic Event," *Proceedings of the Royal Society B: Biological Sciences* 251, no. 1331 (1993): 111–117.

4. Rutz, Christian, and James J. H. St Clair, "The Evolutionary Origins and Ecological Context of Tool Use in New Caledonian Crows," *Behavioural Processes* 89, no. 2 (2012): 153–165.

5. Marzluff, John, and Tony Angell, *Gifts of the Crow: How Perception, Emotion, and Thought Allow Smart Birds to Behave Like Humans* (Free Press, 2012).

6. Mayr, Ernst, "Taxonomic Categories in Fossil Hominids," in *Cold Spring Harbor Symposia on Quantitative Biology*, vol. 15 (Cold Spring Harbor Laboratory Press, 1950), 109–118.

7. Dillard, Annie, "Living Like Weasels," in *Teaching a Stone to Talk: Expeditions and Encounters* (HarperPerennial, 1988), last paragraph.

8. Sol, Daniel, Richard P. Duncan, Tim M. Blackburn, Phillip Cassey, and Louis Lefebvre, "Big Brains, Enhanced Cognition, and Response of Birds to Novel Environments," *Proceedings of the National Academy of Sciences* 102, no. 15 (2005): 5460–5465.

9. Fristoe, Trevor S., and Carlos A. Botero, "Alternative Ecological Strategies Lead to Avian Brain Size Bimodality in Variable Habitats," *Nature Communications* 10, no. 1 (2019): 1–9.

10. Schuck-Paim, Cynthia, Wladimir J. Alonso, and Eduardo B. Ottoni, "Cognition in an Ever-Changing World: Climatic Variability Is Associated with Brain Size in Neotropical Parrots," *Brain, Behavior and Evolution* 71, no. 3 (2008): 200–215.

11. Wagnon, Gigi S., and Charles R. Brown, "Smaller Brained Cliff Swallows Are More Likely to Die During Harsh Weather," *Biology Letters* 16, no. 7 (2020): 20200264.

12. Vincze, Orsolya, "Light Enough to Travel or Wise Enough to Stay? Brain Size Evolution and Migratory Behavior in Birds," *Evolution* 70, no. 9 (2016): 2123–2133.

13. Sayol, Ferran, Joan Maspons, Oriol Lapiedra, Andrew N. Iwaniuk, Tamás Székely, and Daniel Sol, "Environmental Variation and the Evolution of Large Brains in Birds," *Nature Communications* 7, no. 1 (2016): 1–8.

14. Weiner, J., *The Beak of the Finch: A Story of Evolution in Our Time* (Knopf, 1994).

15. Marzluff and Angell, *Gifts of the Crow*, 13.

16. Fristoe, Trevor S., Andrew N. Iwaniuk, and Carlos A. Botero, "Big Brains Stabilize Populations and Facilitate Colonization of Variable Habitats in Birds," *Nature Ecology and Evolution* 1, no. 11 (2017): 1706–1715.

17. Sol, D., J. Maspons, M. Vall-Llosera, I. Bartomeus, G. E. Garcia-Pena, J. Piñol, and R. P. Freckleton, "Unraveling the Life History of Successful Invaders," *Science* 337, no. 6094 (2012): 580–583.

18. Sayol, Ferran, Daniel Sol, and Alex L. Pigot, "Brain Size and Life History Interact to Predict Urban Tolerance in Birds," *Frontiers in Ecology and Evolution* 8 (2020): 58.

19. Oliver, Mary, *New and Selected Poems: Volume One* (Beacon Press, 1992), 220, Kindle.

20. Haupt, Lyanda Lynn, *Crow Planet: Essential Wisdom from the Urban Wilderness* (Little, Brown, 2009).

21. Thoreau, Henry David, *The Journal 1837–1861*, Journal 7, September 1, 1854–October 30, 1855 (New York Review of Books Classics, 2009), chap. 5, January 12, 1855.

22. Sington, David, and Christopher Riley, *In the Shadow of the Moon* (Vertigo Films, 2007), film.

23. Pimm, Stuart L., Julie L. Lockwood, Clinton N. Jenkins, John L. Curnutt, M. Philip Nott, Robert D. Powell, and Oron L. Bass Jr., "Sparrow in the Grass: A Report on the First Ten Years of Research on the Cape Sable Seaside Sparrow (*Ammodramus maritimus mirabilis*)" (unpublished report, 2002), www.nps.gov/ever/learn/nature/upload/MON97-8FinalReportSecure.pdf.

24. Lopez, Barry, *Of Wolves and Men* (Simon and Schuster, 1978).

25. Ducatez, Simon, Daniel Sol, Ferran Sayol, and Louis Lefebvre, "Behavioural Plasticity Is Associated with Reduced Extinction Risk in Birds," *Nature Ecology and Evolution* 4, no. 6 (2020): 788–793.

26. Sol, Daniel, Sven Bacher, Simon M. Reader, and Louis Lefebvre, "Brain Size Predicts the Success of Mammal Species Introduced into Novel Environments," *American Naturalist* 172, no. S1 (2008): S63–S71.

27. Van Woerden, Janneke T., Erik P. Willems, Carel P. van Schaik, and Karin Isler, "Large Brains Buffer Energetic Effects of Seasonal Habitats in Catarrhine Primates," *Evolution: International Journal of Organic Evolution* 66, no. 1 (2012): 191–199.

28. Kalan, Ammie K., et al., "Environmental Variability Supports Chimpanzee Behavioural Diversity," *Nature Communications* 11, no. 1 (2020): 1–10.

29. Marzluff and Angell, *Gifts of the Crow*, 6.

30. Nowell, Branda, and Joseph Stutler, "Public Management in an Era of the Unprecedented: Dominant Institutional Logics as a Barrier to Organizational Sensemaking," *Perspectives on Public Management and Governance* 3, no. 2 (2020): 125–139.

31. Antonson, Nicholas D., Dustin R. Rubenstein, Mark E. Hauber, and Carlos A. Botero, "Ecological Uncertainty Favours the Diversification of Host Use in Avian Brood Parasites," *Nature Communications* 11, no. 1 (2020): 1–7.

32. Beecher, as quoted in the outstanding book by Marzluff, John M., and Tony Angell, *In the Company of Crows and Ravens* (Yale University Press, 2007).

33. Thank you to Clinton Jenkins, Carlos Botero, Branda Nowell, Ferran Sayol, Daniel Sol, Tabby Fenn, Julie Lockwood, Ammie Kalan, John Marzluff, Trevor Brestoe, and Karen Isler for thoughtful comments on this chapter.

Chapter 7: Embracing Diversity to Balance Risk

1. Dillard, Annie, "Life on the Rocks: The Galápagos," section 2, in *Teaching a Stone to Talk: Expeditions and Encounters* (HarperPerennial, 1988).

2. Hutchinson, G. Evelyn, "The Paradox of the Plankton," *American Naturalist* 95, no. 882 (1961): 137–145.

3. Titman, D., "Ecological Competition Between Algae: Experimental Confirmation of Resource-Based Competition Theory," *Science* 192, no. 4238 (1976): 463–465. (Note: this paper was written before David Tilman changed his last name to Tilman.)

4. Tilman, D., and J. A. Downing, "Biodiversity and Stability in Grasslands," *Nature* 367, no. 6461 (1994): 363–365.

5. Tilman, D., P. B. Reich, and J. M. Knops, "Biodiversity and Ecosystem Stability in a Decade-Long Grassland Experiment," *Nature* 441, no. 7093 (2006): 629–632.

6. Dolezal, Jiri, Pavel Fibich, Jan Altman, Jan Leps, Shigeru Uemura, Koichi Takahashi, and Toshihiko Hara, "Determinants of Ecosystem Stability in a Diverse Temperate Forest," *Oikos* 129, no. 11 (2020): 1692–1703.

7. See, for example, Gonzalez, Andrew, et al., "Scaling-Up Biodiversity-Ecosystem Functioning Research," *Ecology Letters* 23, no. 4 (2020): 757–776.

8. Cadotte, Marc W., "Functional Traits Explain Ecosystem Function Through Opposing Mechanisms, *Ecology Letters* 20, no. 8 (2017): 989–996.

9. Martin, Adam R., Marc W. Cadotte, Marney E. Isaac, Rubén Milla, Denis Vile, and Cyrille Violle, "Regional and Global Shifts in Crop Diversity Through the Anthropocene," *PLOS ONE* 14, no. 2 (2019): e0209788.

10. Khoury, Colin K., Anne D. Bjorkman, Hannes Dempewolf, Julian Ramirez-Villegas, Luigi Guarino, Andy Jarvis, Loren H. Rieseberg, and Paul C. Struik, "Increasing Homogeneity in Global Food Supplies and the Implications for Food Security," *Proceedings of the National Academy of Sciences* 111, no. 11 (2014): 4001–4006.

11. Mitchell, Charles E., David Tilman, and James V. Groth, "Effects of Grassland Plant Species Diversity, Abundance, and Composition on Foliar Fungal Disease," *Ecology* 83, no. 6 (2002): 1713–1726.

12. Khoury et al., "Increasing Homogeneity in Global Food Supplies and the Implications for Food Security."

13. Zhu, Youyong, et al., "Genetic Diversity and Disease Control in Rice," *Nature* 406, no. 6797 (2000): 718–722.

14. Bowles, Timothy M., et al., "Long-Term Evidence Shows That Crop-Rotation Diversification Increases Agricultural Resilience to Adverse Growing Conditions in North America," *One Earth* 2, no. 3 (2020): 284–293.

15. Thanks to Marc Cadotte, Nick Haddad, Colin Khoury, Matthew Booker, Stan Harpole, and Nate Sanders for excellent comments and insights on this chapter. Delphine Renard patiently helped me through multiple versions of this chapter.

Chapter 8: The Law of Dependence

1. "Safe Prevention of the Primary Cesarean Delivery," *Obstetric Care Consensus*, no. 1 (2014), https://web.archive.org/web/20140302063757/http://www.acog.org/Resources_And_Publications/Obstetric_Care_Consensus_Series/Safe_Prevention_of_the_Primary_Cesarean_Delivery.

2. Neut, C., et al., "Bacterial Colonization of the Large Intestine in Newborns Delivered by Cesarean Section," *Zentralblatt für Bakteriologie, Mikrobiologie und Hygiene. Series A: Medical Microbiology, Infectious Diseases, Virology, Parasitology* 266, nos. 3–4 (1987): 330–337; Biasucci, Giacomo, Belinda Benenati, Lorenzo Morelli, Elena Bessi, and Günther Boehm, "Cesarean Delivery May Affect the Early Biodiversity of Intestinal Bacteria," *Journal of Nutrition* 138, no. 9 (2008): 1796S–1800S.

3. Leidy, Joseph, *Parasites of the Termites* (Collins, printer, 1881), 425.

4. Tung, Jenny, Luis B. Barreiro, Michael B. Burns, Jean-Christophe Grenier, Josh Lynch, Laura E. Grieneisen, Jeanne Altmann, Susan C. Alberts, Ran Blekhman, and Elizabeth A. Archie, "Social Networks Predict Gut Microbiome Composition in Wild Baboons," *elife* 4 (2015): e05224.

5. Dunn, Robert R., Katherine R. Amato, Elizabeth A. Archie, Mimi Arandjelovic, Alyssa N. Crittenden, and Lauren M. Nichols, "The Internal, External and Extended Microbiomes of Hominins," *Frontiers in Ecology and Evolution* 8 (2020): 25.

6. Godoy-Vitorino, Filipa, Katherine C. Goldfarb, Eoin L. Brodie, Maria A. Garcia-Amado, Fabian Michelangeli, and Maria G. Domínguez-Bello, "Developmental Microbial Ecology of the Crop of the Folivorous Hoatzin," *ISME Journal* 4, no. 5 (2010): 611–620; Godoy-Vitorino, Filipa, Katherine C. Goldfarb, Ulas Karaoz, Sara Leal, Maria A. Garcia-Amado, Philip Hugenholtz, Susannah G. Tringe, Eoin L. Brodie, and Maria Gloria Dominguez-Bello, "Comparative Analyses of Foregut and Hindgut Bacterial Communities in Hoatzins and Cows," *ISME Journal* 6, no. 3 (2012): 531–541.

7. Escherich, T., "The Intestinal Bacteria of the Neonate and Breast-Fed Infant," *Clinical Infectious Diseases* 10, no. 6 (1988): 1220–1225.

8. Domínguez-Bello, Maria G., Elizabeth K. Costello, Monica Contreras, Magda Magris, Glida Hidalgo, Noah Fierer, and Rob Knight, "Delivery Mode Shapes the Acquisition and Structure of the Initial Microbiota Across Multiple Body Habitats in Newborns," *Proceedings of the National Academy of Sciences* 107, no. 26 (2010): 11971–11975.

9. Montaigne, Michel de, *In Defense of Raymond Sebond* (Ungar, 1959).

10. Mitchell, Caroline, et al., "Delivery Mode Affects Stability of Early Infant Gut Microbiota," *Cell Reports Medicine* 1, no. 9 (2020): 100156.

11. Song, Se Jin, et al., "Cohabiting Family Members Share Microbiota with One Another and with Their Dogs," *elife* 2 (2013): e00458.

12. Beasley, D. E., A. M. Koltz, J. E. Lambert, N. Fierer, and R. R. Dunn, "The Evolution of Stomach Acidity and Its Relevance to the Human Microbiome," *PLOS ONE* 10, no. 7 (2015): e0134116.

13. Arboleya, Silvia, Marta Suárez, Nuria Fernández, L. Mantecón, Gonzalo Solís, M. Gueimonde, and C. G. de Los Reyes-Gavilán, "C-Section and the Neonatal Gut Microbiome Acquisition: Consequences for Future Health," *Annals of Nutrition and Metabolism* 73, no. 3 (2018): 17–23.

14. Degnan, Patrick H., Adam B. Lazarus, and Jennifer J. Wernegreen, "Genome Sequence of *Blochmannia pennsylvanicus* Indicates Parallel Evolutionary Trends Among Bacterial Mutualists of Insects," *Genome Research* 15, no. 8 (2005): 1023–1033.

15. Fan, Yongliang, and Jennifer J. Wernegreen, "Can't Take the Heat: High Temperature Depletes Bacterial Endosymbionts of Ants," *Microbial Ecology* 66, no. 3 (2013): 727–733.

16. Lopez, Barry, *Of Wolves and Men* (Simon and Schuster, 1978), chap. 1, "Origin and Description."

17. Maria Gloria Dominguez-Bello, Michael Poulsen, Aram Mikaelyan, Jiri Hulcr, Christine Nalepa, Sandra Breum Andersen, Elizabeth Costello, Jennifer Wernegreen, Noah Fierer, and Filipa Godoy-Vitorino all provided insightful comments on this chapter. Thank you.

Chapter 9: Humpty-Dumpty and the Robotic Sex Bees

1. Tsui, Clement K.-M., Ruth Miller, Miguel Uyaguari-Diaz, Patrick Tang, Cedric Chauve, William Hsiao, Judith Isaac-Renton, and Natalie Prystajecky, "Beaver Fever: Whole-Genome Characterization of Waterborne Outbreak and Sporadic Isolates to Study the Zoonotic Transmission of Giardiasis," *mSphere* 3, no. 2 (2018): e00090-18.

2. McMahon, Augusta, "Waste Management in Early Urban Southern Mesopotamia," in *Sanitation, Latrines and Intestinal Parasites in Past Populations*, ed. Piers D. Mitchell (Farnham, 2015), 19–40.

3. National Research Council, *Watershed Management for Potable Water Supply: Assessing the New York City Strategy* (National Academies Press, 2000).

4. Gebert, Matthew J., Manuel Delgado-Baquerizo, Angela M. Oliverio, Tara M. Webster, Lauren M. Nichols, Jennifer R. Honda, Edward D. Chan, Jennifer Adjemian, Robert R. Dunn, and Noah Fierer, "Ecological Analyses of Mycobacteria in Showerhead Biofilms and Their Relevance to Human Health," *MBio* 9, no. 5 (2018).

5. Proctor, Caitlin R., Mauro Reimann, Bas Vriens, and Frederik Hammes, "Biofilms in Shower Hoses," *Water Research* 131 (2018): 274–286.

6. For more on this research, see a longer discussion in Dunn, Rob, *Never Home Alone: From Microbes to Millipedes, Camel Crickets, and Honeybees, the Natural History of Where We Live* (Basic Books, 2018).

7. Ngor, Lyna, Evan C. Palmer-Young, Rodrigo Burciaga Nevarez, Kaleigh A. Russell, Laura Leger, Sara June Giacomini, Mario S. Pinilla-Gallego, Rebecca E. Irwin, and Quinn S. McFrederick, "Cross-Infectivity of Honey and Bumble Bee–Associated Parasites Across Three Bee Families," *Parasitology* 147, no. 12 (2020): 1290–1304.

8. Knops, Johannes M. H., et al., "Effects of Plant Species Richness on Invasion Dynamics, Disease Outbreaks, Insect Abundances and Diversity," *Ecology Letters* 2, no. 5 (1999): 286–293.

9. Tarpy, David R., and Thomas D. Seeley, "Lower Disease Infections in Honeybee (*Apis mellifera*) Colonies Headed by Polyandrous vs Monandrous Queens," *Naturwissenschaften* 93, no. 4 (2006): 195–199.

10. Zattara, Eduardo E., and Marcelo A. Aizen, "Worldwide Occurrence Records Suggest a Global Decline in Bee Species Richness," *One Earth* 4, no. 1 (2021): 114–123.

11. Potts, S. G., P. Neumann, B. Vaissière, and N. J. Vereecken, "Robotic Bees for Crop Pollination: Why Drones Cannot Replace Biodiversity," *Science of the Total Environment* 642 (2018): 665–667.

12. Thank you to David Tarpy, Charles Mitchell, Angela Harris, Nicolas Vereecken, Brad Taylor, Becky Irwin, Kendra Brown, Margarita Lopez Uribe, and Noah Fierer for comments on this chapter.

Chapter 10: Living with Evolution

1. Warner Bros. Canada, "Contagion: Bacteria Billboard," September 7, 2011, YouTube video, 1:38, www.youtube.com/watch?v=LppK4ZtsDdM&feature=emb_title.

2. Weiner, J., *The Beak of the Finch: A Story of Evolution in Our Time* (Knopf, 1994), 9.

3. Darwin, Charles, *The Descent of Man*, 6th ed. (Modern Library, 1872), chap. 4, fifth paragraph.

4. Fleming, Sir Alexander, "Banquet Speech," December 10, 1945, The Nobel Prize, www.nobelprize.org/prizes/medicine/1945/fleming/speech/.

5. Comte de Buffon, Georges-Louis Leclerc, *Histoire naturelle, générale et particulière*, vol. 12, *Contenant les époques de la nature* (De l'Imprimerie royale, 1778), 197.

6. Jørgensen, Peter Søgaard, Carl Folke, Patrik J. G. Henriksson, Karin Malmros, Max Troell, and Anna Zorzet, "Coevolutionary Governance of Antibiotic and Pesticide Resistance," *Trends in Ecology and Evolution* 35, no. 6 (2020): 484–494.

7. Aktipis, Athena, "Applying Insights from Ecology and Evolutionary Biology to the Management of Cancer, an Interview with Athena Aktipis," interview by Rob Dunn, *Applied Ecology News*, July 28, 2020, https://cals.ncsu.edu/applied-ecology/news/ecology-and-evolutionary-biology-to-the-management-of-cancer-athena-aktipis/.

8. Harrison, Freya, Aled E. L. Roberts, Rebecca Gabrilska, Kendra P. Rumbaugh, Christina Lee, and Stephen P. Diggle, "A 1,000-Year-Old Antimicrobial Remedy with Antistaphylococcal Activity," *mBio* 6, no. 4 (2015): e01129-15.

9. Aktipis, Athena, *The Cheating Cell: How Evolution Helps Us Understand and Treat Cancer* (Princeton University Press, 2020).

10. Jørgensen, Peter S., Didier Wernli, Scott P. Carroll, Robert R. Dunn, Stephan Harbarth, Simon A. Levin, Anthony D. So, Maja Schlüter, and Ramanan Laxminarayan, "Use Antimicrobials Wisely," *Nature* 537, no. 7619 (2016): 159.

11. I'm grateful to Peter Jørgensen, Athena Aktipis, Michael Baym, Roy Kishony, Tami Lieberman, and Christina Lee for their thoughtful comments on and additions to this chapter.

Chapter 11: Not the End of Nature

1. Dunn, Robert R., "Modern Insect Extinctions, the Neglected Majority," *Conservation Biology* 19, no. 4 (2005): 1030–1036.

2. Koh, Lian Pin, Robert R. Dunn, Navjot S. Sodhi, Robert K. Colwell, Heather C. Proctor, and Vincent S. Smith, "Species Coextinctions and the Biodiversity Crisis," *Science* 305, no. 5690 (2004): 1632–1634. See also a later summary of this approach in Dunn, Robert R., Nyeema C. Harris, Robert K. Colwell, Lian Pin Koh, and Navjot S. Sodhi, "The Sixth Mass Coextinction: Are Most Endangered Species Parasites and Mutualists?," *Proceedings of the Royal Society B: Biological Sciences* 276, no. 1670 (2009): 3037–3045.

3. Pimm, Stuart L., *The World According to Pimm: A Scientist Audits the Earth* (McGraw-Hill, 2001).

4. As Nee put it in a later book chapter based on the talk, big species "jump around and make a lot of noise," but they "represent very little of biological diversity." By "big," he meant things the size of a mite and larger, from mites to moose. Nee, Sean, "Phylogenetic Futures After the Latest Mass Extinction," in *Phylogeny and Conservation*, ed. Purvis, Andrew, John L. Gittleman, and Thomas Brooks (Cambridge University Press, 2005), 387–399.

5. Jenkins, Clinton N., et al., "Global Diversity in Light of Climate Change: The Case of Ants," *Diversity and Distributions* 17, no. 4 (2011): 652–662.

6. Wehner, Rüdiger, *Desert Navigator: The Journey of an Ant* (Harvard University Press, 2020), 25.

7. Willot, Quentin, Cyril Gueydan, and Serge Aron, "Proteome Stability, Heat Hardening and Heat-Shock Protein Expression Profiles in *Cataglyphis* Desert Ants," *Journal of Experimental Biology* 220, no. 9 (2017): 1721–1728.

8. Perez, Rémy, and Serge Aron, "Adaptations to Thermal Stress in Social Insects: Recent Advances and Future Directions," *Biological Reviews* 95, no. 6 (2020): 1535–1553.

9. Nesbitt, Lewis Mariano, *Hell-Hole of Creation: The Exploration of Abyssinian Danakil* (Knopf, 1935), 8.

10. Gómez, Felipe, Barbara Cavalazzi, Nuria Rodríguez, Ricardo Amils, Gian Gabriele Ori, Karen Olsson-Francis, Cristina Escudero, Jose M. Martínez, and Hagos Miruts, "Ultra-Small Microorganisms in the

Polyextreme Conditions of the Dallol Volcano, Northern Afar, Ethiopia," *Scientific Reports* 9, no. 1 (2019): 1–9.

11. Cavalazzi, B., et al., "The Dallol Geothermal Area, Northern Afar (Ethiopia)—An Exceptional Planetary Field Analog on Earth," *Astrobiology* 19, no. 4 (2019): 553–578.

12. I'm very grateful to Felipe Gómez, Barbara Cavalazzi, Robert Colwell, Mary Schweitzer, Russell Lande, Jamie Shreeve, Serge Aron, Xim Cerda, Cat Cardelus, Clinton Jenkins, Lian Pin Koh, and Sean Nee for reading this chapter and providing their insights. Thank you, Laura Hug, for your remarkable phylogeny.

Conclusion

1. Marshall, Charles R., "Five Palaeobiological Laws Needed to Understand the Evolution of the Living Biota," *Nature Ecology and Evolution* 1, no. 6 (2017): 1–6.

2. Hagen, Oskar, Tobias Andermann, Tiago B. Quental, Alexandre Antonelli, and Daniele Silvestro, "Estimating Age-Dependent Extinction: Contrasting Evidence from Fossils and Phylogenies," *Systematic Biology* 67, no. 3 (2018): 458–474.

3. Harris, Nyeema C., Travis M. Livieri, and Robert R. Dunn, "Ectoparasites in Black-Footed Ferrets (*Mustela nigripes*) from the Largest Reintroduced Population of the Conata Basin, South Dakota, USA," *Journal of Wildlife Diseases* 50, no. 2 (2014): 340–343.

4. Colwell, Robert K., Robert R. Dunn, and Nyeema C. Harris, "Coextinction and Persistence of Dependent Species in a Changing World," *Annual Review of Ecology, Evolution, and Systematics* 43 (2012):183–203.

5. Rettenmeyer, Carl W., M. E. Rettenmeyer, J. Joseph, and S. M. Berghoff, "The Largest Animal Association Centered on One Species: The Army Ant *Eciton burchellii* and Its More Than 300 Associates," *Insectes Sociaux* 58, no. 3 (2011): 281–292.

6. Penick, Clint A., Amy M. Savage, and Robert R. Dunn, "Stable Isotopes Reveal Links Between Human Food Inputs and Urban Ant Diets," *Proceedings of the Royal Society B: Biological Sciences* 282, no. 1806 (2015): 20142608.

7. Dunn, Robert R., Charles L. Nunn, and Julie E. Horvath, "The Global Synanthrome Project: A Call for an Exhaustive Study of Human Associates," *Trends in Parasitology* 33, no. 1 (2017): 4–7.

8. Panagiotakopulu, Eva, Peter Skidmore, and Paul Buckland, "Fossil Insect Evidence for the End of the Western Settlement in Norse Greenland," *Naturwissenschaften* 94, no. 4 (2007): 300–306.

9. Weisman, Alan, *The World Without Us* (Macmillan, 2007), 8.

10. Marshall, "Five Palaeobiological Laws Needed to Understand the Evolution of the Living Biota."

11. Losos, Jonathan B., *Improbable Destinies: Fate, Chance, and the Future of Evolution* (Riverhead Books, 2017).

12. Hoekstra, Hopi E., "Genetics, Development and Evolution of Adaptive Pigmentation in Vertebrates," *Heredity* 97, no. 3 (2006): 222–234.

13. Sayol, F., M. J. Steinbauer, T. M. Blackburn, A. Antonelli, and S. Faurby, "Anthropogenic Extinctions Conceal Widespread Evolution of Flightlessness in Birds," *Science Advances* 6, no. 49 (2020): eabb6095.

14. Losos, Jonathan B., *Lizards in an Evolutionary Tree: Ecology and Adaptive Radiation of Anoles* (University of California Press, 2011).

15. Braude, Stanton, "The Predictive Power of Evolutionary Biology and the Discovery of Eusociality in the Naked Mole-Rat," *Reports of the National Center for Science Education* 17, no. 4 (1997): 12–15.

16. Jarvis, J. U., "Eusociality in a Mammal: Cooperative Breeding in Naked Mole-Rat Colonies," *Science* 212, no. 4494 (1981): 571–573; Sherman, Paul W., Jennifer U. M. Jarvis, and Richard D. Alexander, eds., *The Biology of the Naked Mole-Rat* (Princeton University Press, 2017).

17. Feigin, C. Y., et al., "Genome of the Tasmanian Tiger Provides Insights into the Evolution and Demography of an Extinct Marsupial Carnivore," *Nature Ecology and Evolution* 2 (2018):182–192.

18. D'Ambrosia, Abigail R., William C. Clyde, Henry C. Fricke, Philip D. Gingerich, and Hemmo A. Abels, "Repetitive Mammalian Dwarfing During Ancient Greenhouse Warming Events," *Science Advances* 3, no. 3 (2017): e1601430.

19. Smith, Felisa A., Julio L. Betancourt, and James H. Brown, "Evolution of Body Size in the Woodrat over the Past 25,000 Years of Climate Change," *Science* 270, no. 5244 (1995): 2012–2014.

20. Zalasiewicz, Jan, and Kim Freedman, *The Earth After Us: What Legacy Will Humans Leave in the Rocks?* (Oxford University Press, 2009).

21. Zalasiewicz and Freedman, *The Earth After Us*, chap. 2, section "Future Earth: Close Up."

22. Losos, Jonathan, "Lizards, Convergent Evolution and Life After Humans, an Interview with Jonathan Losos," interview by Rob Dunn,

Applied Ecology News, September 21, 2020, https://cals.ncsu.edu/applied-ecology/news/lizards-convergent-evolution-and-life-after-humans-an-interview-with-jonathan-losos/.

23. Gould, Stephen Jay, *Full House* (Harvard University Press, 1996), 176.

24. Thank you to Bucky Gates, Lindsay Zanno, Jan Zalasiewicz, Mary Schweitzer, Jonathan Losos, Charles Marshall, Robert Colwell, Christy Hipsley, Alan Weisman, Tom Gilbert, Eva Panagiotakopulu, and Lian Pin Koh for reading this chapter and offering their insights and expertise.

Index

ROB DUNN is Reynolds Professor in the Department of Applied Ecology at North Carolina State University and a professor of human biodiversity at the Center for Evolutionary Hologenomics at the University of Copenhagen. He is the author of six books.